ESTIMATES FOR THE $\bar{\partial}$-NEUMANN PROBLEM

P. C. GREINER and E. M. STEIN

Princeton University Press
1977

Published by Princeton University Press, Princeton, New Jersey

In the United Kingdom: Princeton University Press, Guildford, Surrey

Library of Congress Cataloging in Publication Data will
be found on the last printed page of this book

Published in Japan exclusively
by University of Tokyo Press
in other parts of the world by
Princeton University Press

Printed in the United States of America
by Princeton University Press, Princeton, New Jersey

Preface

The $\bar{\partial}$-Neumann problem is probably the most important and natural example of a non-elliptic boundary value problem, arising as it does from the Cauchy-Riemann equations. It has been known for some time, going back to the work of Kohn, how to determine solvability and prove regularity of solutions in the study of this problem. The main tool that was used were various L^2 estimates. What had been lacking was a more explicit construction of the operator (or at least an approximation) giving the solution, and the study of estimates in other norms, all of which would be needed in order to have a better understanding of the structure of the $\bar{\partial}$-Neumann problem

It is our primary purpose here to describe the work we have done in this direction in the last few years.* We construct parametricies and give sharp estimates in appropriate function spaces.

Our secondary aim in writing this monograph is didactic. We present not only the required details of the proofs of our main results, but also certain prerequisites and some additional material which we hope will give the reader a clearer view of the whole subject.

It is with pleasure that we take this opportunity to express our appreciation to Miss Florence Armstrong for her excellent job of typing the manuscript.

November 1976

---

*A preliminary announcement of our main results was given in the 1974 Montreal Conference [12].

Table of Contents

Introduction

Why the $\bar{\partial}$-Neumann problem?

Let M be an open relatively compact subset of a complex manifold M' of dimension $n+1$, and assume that the boundary of M, bM, is smooth and strongly pseudo-convex. The $\bar{\partial}$-Neumann problem for M arises when one tries to solve the Cauchy-Riemann equations on M

$$(1) \qquad \bar{\partial}U = f,$$

where f is a given (0,1) form with

$$(2) \qquad \bar{\partial}f = 0,$$

and in particular when one wishes to have control on the behavior of U near the boundary in terms of similar control over f.

Part of the difficulty of this problem is connected with the fact that (1) is overdetermined, and also without some further conditions the solution of (1) is not uniquely specified. There is however a general formalism due to Spencer (applicable also in many other situations) which gets around these initial difficulties. Instead of (1) and (2) one considers the second-order equation

$$(3) \qquad \Box u = (\bar{\partial}\bar{\partial}^* + \bar{\partial}^*\bar{\partial})u = f.$$

$\bar{\partial}^*$ is the adjoint of $\bar{\partial}$ (which is defined once one chooses a fixed Hermitian metric on $\bar{M}$). To explain (3) more precisely we temporarily write (1) as $\bar{\partial}_0 U = f$, and (2) as $\bar{\partial}_1 f = 0$, where the subscripts indicate that the first $\bar{\partial}$ acts on functions, and the second $\bar{\partial}$ on (0,1) forms. Strictly speaking $\square$ should be written as $\bar{\partial}_0 \bar{\partial}_0^* + \bar{\partial}_1^* \bar{\partial}_1$. And now it is clear that the equation (3) comes naturally equipped with a pair of boundary conditions

(4) $\quad u \in \text{domain}\ (\bar{\partial}_0^*)$

(5) $\quad \bar{\partial}_1 u \in \text{domain}\ (\bar{\partial}_1^*)$.

The equation (3), which is essentially Laplace's equation, together with the two boundary conditions (4) and (5) give us the $\bar{\partial}$-Neumann problem. It can be shown that if u is a solution of this problem, with $\bar{\partial}f = 0$, then $U = \bar{\partial}^* u$ solves our original problem (1), and is in addition uniquely specified by the property that U is orthogonal to holomorphic functions on $\overline{M}$

Kohn's solution, and some further problems

The analytic difficulty of the problem is due to the fact that while the differential operator $\square$ is elliptic, the boundary conditions are not. Nevertheless, using L^2 estimates, Kohn was able to prove existence and make a systematic analysis of the regularity properties of solutions of this problem. One of his main results is the estimate

(6) $$\|u\|_{L^2_{k+1}} \leq A_k(\|\square u\|_{L^2_k} + \|u\|_{L^2}), \qquad n=0,1,2,\ldots$$

for u satisfying the boundary conditions (4) and (5). Thus there was a gain of 1, and not 2 as in the standard elliptic boundary value problems.

The problems that were left open were as follows:

(i) To understand more fully these regularity properties, even in the context of the L^2 norm,

(ii) To find the corresponding estimates for other function spaces, e.g., L^p spaces, Lipschitz spaces, etc.,

(iii) To give a more explicit construction of the operator (the Neumann operator), which expresses the solution u in terms of f.

To deal with (i) - (iii) is the main task of this monograph.*

Three principles

We shall be guided by three principles, the first of which is by now well-understood.

First, the solution of a boundary-value problem for a differential operator which is elliptic can, by the use of the theory of pseudo-differential operators, be reduced to the inversion of a pseudo-differential operator acting on the boundary. So our first task is to isolate this pseudo-differential operator. This is the operator $\square^+$ (see Chapter 8), and since it is not elliptic we need to know it rather precisely: $\square^+$ is of order 1, but its zero order terms are not negligible.

In inverting $\square^+$ we are guided by the second principle, namely that the inverse of $\square^+$ should be modeled on the inverse of $\square^+$ in the special case corresponding to the Heisenberg group. This accounts for the key role of the Heisenberg group in our analysis, which incidentally

*Background material that might be useful for the reading of this monograph is contained in the survey [35].

is closely related to the similar role it plays in the case of the boundary analogue of (1) or (3).*

The third principle follows from the second. All estimates which are sharp will reflect the structure of the Heisenberg group (or what amounts to the same thing, the complex structure of M). Thus there are "good" directions which are singled out, and in terms of these directions we have a gain of two, as in the usual elliptic case. Let us now describe these things in greater detail.

Reduction to a boundary problem

We first solve the Dirichlet boundary-value problem for $\square$, and our procedure here uses well-known techniques. We construct Green's operator G which solves the inhomogeneous problem with zero Dirichlet boundary conditions, and the Poisson integral P, which solves the homogeneous problem with given Dirichlet boundary conditions. Thus

(7) $$u = G(\square u) + P(u_b)$$

where u_b is the restriction of u to the boundary.
Actually we only construct an approximate version of (7), valid in appropriate coordinate patches near the boundary (see Theorem 7.66), but this suffices for our purposes. Because of the non-elliptic nature of the boundary-value problem for which we want to use (7), it is crucial that we keep track of the symbols one order less than the top order,

*See Folland and Stein [9].

and this makes our calculations somewhat elaborate.

Now let $B_{\bar\partial}$ be the boundary operator giving the second $\bar\partial$-Neumann boundary condition (i.e., (5)). Then the basic boundary operator we have to deal with (and whose symbol we determine rather precisely) is

(8) $$\Box^+ = B_{\bar\partial} P$$

Inversion of $\Box^+$

Using the symbolic calculus we find another operator, $\Box^-$, in many ways similar to $\Box^+$, so that

(9) $$\Box_b = -\Box^- \Box^+ \quad \text{approximately.}$$

Here $\Box_b$ is the Kohn Laplacian for the $\bar\partial_b$ complex (acting on (0,1) forms on bM).

Now when $n > 1$, $\Box_b$ has an approximate inverse K, given as an integral operator modeled on a convolution operator on the Heisenberg group. (This is one of the main results of the paper [9]). Thus when $n > 1$ an approximate inverse to $\Box^+$ is $-K\Box^-$; with this it is a straightforward matter to write down an approximation to the Neumann operator, (giving the solution to the $\bar\partial$-Neumann problem) in terms of $K\Box^-$, $B_{\bar\partial}$, P and G. (See Proposition 9.26.)

The case n=1

When n=1 the operator $\Box_b$ is not invertible, and therefore a further analysis is required. The idea is as follows. Near the characteristic variety of $\Box^+$, the operator $\Box_b$ behaves in the case n=1 like it does in the case $n > 1$. Hence near the characteristic variety of $\Box^+$ we can write an inverse of $\Box^+$ similar to that for $n > 1$. However, away from its characteristic variety, $\Box^+$ is elliptic, and so here we can find an inverse by the use of the standard calculus of pseudo-differential operators.

Now the required analysis for $\Box_b$ (when $n=1$) can be done in several ways. The most elegant approach is via the identity

$$\overline{K}\,\Box_b = \Box_b\,\overline{K} = I - \overline{C}_b \tag{10}$$

on the Heisenberg group which was obtained in a joint work [11] with Kohn.*

Here $\overline{K}$ is an (explicit) convolution operator (of type 2) in the Heisenberg group, and C_b is the Cauchy-Szegö projection. Incidentally the identity (10) leads to the necessary and sufficient conditions for local solvability of $\Box_b$ when $n=1$, for $\Box_b$ on functions for any n, and for the Lewy operator, when $n=1$. (For further details see Chapter 3.)

The inverse of $\Box^+$ (when $n=1$) can then be given two alternative (but roughly equivalent) forms in terms of $\overline{K}$ and $\overline{C}_b$; see Lemma 10.25, and Lemma 10.32.

The estimates

Once the analysis of the inverse of $\Box^+$ is concluded, the estimates for the solution of the $\bar{\partial}$-Neumann problem (i.e., control of the Neumann

*Our original approach, sketched in [12], was more complicated. It used the material in Chapters 1 and 2.

operator) can be given in terms of corresponding estimates for four classes of operators: The restriction operator, standard pseudo-differential operators, convolution operators of the Heisenberg type studied in [9], and Poisson operators. For the first two classes of operators the estimates we need are known. For the other two classes we need to invoke the results of [9] (see also [29]), and some new estimates need to be made. The required work is done in Chapters 12 to 15.

Our main conclusions are then as follows. Suppose u is the solution of the problem (3) with boundary condition (4) and (5). Suppose $f \in L^p_k(\overline{M})$, for some p, $1 < p < \infty$, with k a non-negative integer. Then (see Theorem 15.1),

(a) $u \in L^p_{k+1}(\overline{M})$; u gains one in all directions.

(b) u gains two in the "good" directions, i.e., $p(Z, \overline{Z})u \in L^p_k(M)$, where p is any polynomial of degree 2 in "allowable" vector fields.

(c) In the anti-holomorphic direction normal to the boundary, the gain is two in the following strong sense. Suppose Z_{n+1} is a holomorphic vector field whose real part is proportional to $\frac{\partial}{\partial \rho}$, where ρ is the distance to the boundary. Then $\overline{Z}_{n+1} u \in L^p_{k+1}(\overline{M})$.*

There are similar results for the Lipschitz spaces given in Theorems 15.30 and 15.33.

*That all anti-holomorphic derivatives of u should be better behaved than holomorphic derivatives, was suggested to us by Kohn. In fact it turns out that this is the case because of the second boundary condition (5).

Estimates for $\bar{\partial}U = f$

With all of this one is finally in a position to obtain the estimates for the solution of (1) which is orthogonal to holomorphic functions. If f satisfies the necessary conditions (i.e., $\bar{\partial}f = 0$, and f orthogonal to the harmonic space*), then when $f \in L^p_k(\overline{M})$ we have:

(a) U gains 1/2 in all directions, i.e., $U \in L^p_{k+1/2}(\overline{M})$

(b) U gains 1 in good directions, i.e., $XU \in L^p_k(\overline{M})$, where X is any allowable vector field.

Similar results hold for the Lipschitz spaces; see Theorem 16.7.

Some open problems

All our analysis is carried out in terms of a fixed Hermitian metric on $\overline{M}$ which is compatible with the Levi-forms on bM (a "Levi metric"). This limitation was imposed because of a similar restriction in [9] for $\Box_b$. However recently, in Rothschild and Stein [29], it is shown that this limitation for $\Box_b$ may be dropped, and so it is very reasonable to suppose that our results for the $\bar{\partial}$-Neumann problem, and for the solution of $\bar{\partial}U = f$, will go through without any restriction on the Hermitian metric.

It also seems highly likely that similar results holds for (0,q) forms generally. Again by the recent results of [29] it is reasonable to suppose that for (0,q) forms our results will hold if the Levi form is no longer assumed to be positive definite but satisfies condition Z(q)

*The harmonic space is empty if M' is a Stein manifold, in particular if M is a sub-domain of $\mathbb{C}^{n+1}$.

of [8], p. 57, namely the Levi form has at least $n+1-q$ positive eigenvalues or at least $q+1$ negative eigenvalues at each point of bM.

A final problem which needs to be solved is as follows:

It would be interesting to know an explicit expression for the kernel or the symbol of the Neumann operator. We have, in effect, obtained this operator as a product of other operators whose form we know quite precisely. However the explicit description of the resulting operator is still somewhat of a mystery.

Part I. Analysis on the Heisenberg group

Guide to Part I

Notations. Let $\mathcal{D}$ be the domain $\{(z_1,\ldots,z_{n+1}),\ \mathrm{Im}\, z_{n+1} > |z_1|^2 + |z_2|^2 \ldots + |z_n|^2\}$.* The subgroup of holomorphic self-mappings of $\mathcal{D}$ which consist of "translations" is called the Heisenberg group H_n. Since it acts simply-transitively on the boundary of $\mathcal{D}$, it can be naturally identified with this boundary.

If $(\zeta, t) \in \mathbb{C}^n \times \mathbb{R}$ denotes a point of H_n, then it is identified with the boundary point given by

(0.1) $$z_{n+1} = t + i|\zeta|^2, \qquad (z_1, z_2, \ldots, z_n) = \zeta$$

The multiplication law on H_n is

(0.2) $$(\zeta, t)\cdot(\zeta', t') = (\zeta+\zeta',\ t+t'+2\,\mathrm{Im}\,\zeta\cdot\bar{\zeta}').$$

Left-invariant vector fields on H_n are given by the Z_j (see (1.1)) which are restrictions to $b\mathcal{D}$ of holomorphic vector fields, tangential at $b\mathcal{D}$. (We use the notation $t = x_0$, $z_j = x_{2j-1} + ix_{2j}$ $j=1,\ldots,n$). The Z_j, $\bar{Z}_j$, and $T = \frac{\partial}{\partial t}$ form a basis for the left-invariant vector fields on H_n. For further background on this, see [9], §§4 and 5, and also [22]

The operators $\mathcal{L}_\alpha$

The operators $\mathcal{L}_\alpha$ (see (1.2)), are essentially the $\Box_b$ Laplacians restricted to q-forms, where $\alpha = n-2q$, (see the discussion in Chapter 3,

*As is well known, the domain $\mathcal{D}$ is holomorphically equivalent to the unit ball in $\mathbb{C}^{n+1}$.

and in particular (3.5)). The main purpose of Part I is to study the inverse to this operator, when it exists.

The symbol of the inverse operator

There are two ways of expressing the inverse to $\mathcal{L}_\alpha$. One is as an integral operator

$$f \longrightarrow \int G_\alpha(x,y)f(y)dy$$

whose kernel G_α is a fundamental solution of $\mathcal{L}_\alpha$. In view of the fact that $\mathcal{L}_\alpha$ is left-invariant, one can take

$$G_\alpha(x,y) = G_\alpha(y^{-1}x, 0). \tag{0.3}$$

An explicit form for G_α was first found in Folland-Stein [9], (see Chapter 2 below, where $G_\alpha(x,0)$ is written as $\Phi_\alpha(x)$). Another expression of the operator is in terms of its symbol, which in view of the multiplication rule (0.2) is essentially the Euclidean Fourier transform of $G_\alpha(x,0)$. We begin by deriving the latter form.

Derivation

The Fourier transform of G_α is studied by first making a partial Fourier transform in the t-variable. After some reduction, one is led to a confluent hypergeometric equation (1.7), for which one can write an integral expression for one of its solutions, (converging when Re $n-\alpha > 0$, $n > 1$ and when the dual variable lies in a half-space). Next one takes the Fourier transform in the other variables, obtaining the symbol of the operator in a half-space, when Re $(n-\alpha) > 0$; see Proposition (1.3).

For other values of α (excluding the singular values where $\frac{n \pm \alpha}{2} = 0, -1, -2, \ldots$)) the full symbol is obtained by analytic continuation, which requires replacing an integral over a segment by a loop integral. The final result is in Theorem 1.21 and its corollary. The fact that this result agrees with the original fundamental solution is expressed in Theorem (2.4) and Corollary (2.22).

The singular case $\alpha = \pm n$

The problem of 1-forms in 2 complex variables leads to the operator $\mathcal{L}_\alpha$, where $\alpha = n-2q = -1$, $n=1$. This situation is closely connected with Lewy's equation. There no longer exists a fundamental solution of $\mathcal{L}_\alpha$, when $\alpha = \pm n$, but one can find a relative fundamental solution involving the Cauchy-Szegö projection. The main identity is given in Lemma (3 18). An alternative method of deriving this identity is indicated in Proposition (3.26). The identity (3.18) leads to necessary and sufficient conditions for the local solvability of $\mathcal{L}_\alpha(u) = f$, and for the Lewy equation.

Chapter I. Symbols on the Heisenberg groups

Let

(1.1) $$Z_j = \frac{\partial}{\partial z_j} + \bar{z}_j \frac{\partial}{\partial t}, \quad j=1,\ldots,n$$

be the usual left invariant vector fields on the Heisenberg group H_n. Following the notation and terminology of Folland-Stein [9] we define

(1.2) $$\mathcal{L}_\alpha = -\frac{1}{2}\sum_{j=1}^{n}(Z_j\bar{Z}_j + \bar{Z}_j Z_j) + i\alpha\frac{\partial}{\partial t}$$

$$= \sum_{j=1}^{n}\Bigg(-\frac{\partial^2}{\partial z_j \partial \bar{z}_j} - |z_j|^2 \frac{\partial^2}{\partial t^2}$$

$$-\frac{\partial}{i\partial t}\left(z_j\frac{\partial}{\partial z_j} - \bar{z}_j\frac{\partial}{\partial \bar{z}_j}\right)\Bigg) + i\alpha\frac{\partial}{\partial t}.$$

The purpose of this chapter is to compute the symbol of the fundamental solution of $\mathcal{L}_\alpha$. Let G_α denote a (presumptive) fundamental solution and set

(1.3) $$G_\alpha(t) = \frac{1}{2\pi}\int_{-\infty}^{\infty} \hat{G}_\alpha(\xi_0)\, e^{it\xi_0}\, d\xi_0.$$

Taking $\mathcal{L}_\alpha$ under the Fourier transform we obtain the operator

(1.4) $$\hat{\mathcal{L}}_\alpha = \sum_{j=1}^{n}\Bigg(-\frac{\partial^2}{\partial z_j \partial \bar{z}_j} + |z_j|^2\,\xi_0^2$$

$$-\xi_0\left(z_j\frac{\partial}{\partial z_j} - \bar{z}_j\frac{\partial}{\partial \bar{z}_j}\right)\Bigg) - \alpha\xi_0.$$

Assume $\xi_0 > 0$. (The case $\xi_0 < 0$ will follow by replacing α by

$-\alpha$.) Since we want G_α to act on the Heisenberg group by convolution we shall try a kernel of the following form

$$(1.5) \qquad \hat{G}_\alpha = e^{\xi_0(z\bar{w} - \bar{z}w)} W(\xi_0 |z-w|^2),$$

where we used the notation

$$z_j = x_{2j-1} + ix_{2j},$$

$$w_j = y_{2j-1} + iy_{2j},$$

$j=1,\ldots,n$, $z\bar{w} - \bar{z}w = \sum_{j=1}^{n} (z_j\bar{w}_j - \bar{z}_j w_j)$

and $|z-w|^2 = \sum_{j=1}^{n} |z_j - w_j|^2$.

The choice (1.5) is dictated by the following considerations. First by (0.3) it suffices to consider the special case when $y = 0$ (when $w = 0$). Next $\hat{\mathscr{L}}_\alpha$ is invariant under unitary linear transformation of the z-variables, and so one may look for a $\hat{G}_\alpha$ which depends only on $|z|$. Also observe that $z_j \frac{\partial}{\partial z_j} - \bar{z}_j \frac{\partial}{\partial \bar{z}_j}$ annihalates any function of $|z|$. Thus we are led to the form (1.5) (when $w = 0$). For general w we then use the group law (0.2) to reduce matters to the case $w = 0$. Next we shall solve

$$\hat{\mathscr{L}}_\alpha \hat{G}_\alpha = 0$$

if $z \neq w$. A bit of algebra yields

$$\hat{\mathscr{L}}_\alpha \hat{G}_\alpha = \xi_0 e^{\xi_0(z\bar{w} - \bar{z}w)}$$

$$\cdot \Big\{ -\xi_0 |z-w|^2 W''(\xi_0|z-w|^2) - nW'(\xi_0|z-w|^2)$$

$$+ \xi_0|z-w|^2 W(\xi_0|z-w|^2) - \alpha W(\xi_0|z-w|^2) \Big\}.$$

We set $\xi_0 |z-w|^2 = x$. Hence we need to solve

(1.6) $$x \frac{d^2 W}{dx^2} + n \frac{dW}{dx} + (-x+\alpha) W = 0$$

if $x \neq 0$. This is a confluent hypergeometric differential equation.* We set

$$W(x) = e^{-x} y(2x)$$

which reduces the equation (1.6) to the following better known form

(1.7) $$uy''(u) + (n-u) y'(u) - \frac{n-\alpha}{2} y(u) = 0$$

if $u > 0$. The identity

$$\left(u \frac{d^2}{du^2} + (c-u) \frac{d}{du} - a\right)\left(e^{-us} s^{a-1} (1+s)^{c-a-1}\right)$$

$$= -\frac{d}{ds} \left(e^{-us} s^{a-1} (1+s)^{c-a-1}\right)$$

implies that

$$\int_0^\infty e^{-us} s^{a-1} (1+s)^{c-a-1} ds$$

is a solution of

$$uy''(u) + (c-u) y'(u) - ay(u) = 0$$

if $u > 0$ and $\operatorname{Re} a > 0$. Therefore we assume that

(1.8) $$W(\xi_0 |z-w|^2)$$
$$= a(\xi_0) e^{-\xi_0 |z-w|^2}$$
$$\cdot \int_0^\infty e^{-2\xi_0 |z-w|^2 s} s^{\frac{n-\alpha}{2} - 1} (1+s)^{\frac{n+\alpha}{2} - 1} ds.$$

* See [6], Chapter 6.

We still need to determine the unknown function $a(\xi_0)$. To this end we note that

$$\lim_{x \to 0} x^{n-1} \int_0^\infty e^{-2xs} s^{\frac{n-\alpha}{2}-1} (1+s)^{\frac{n+\alpha}{2}-1} ds$$

$$= \int_0^\infty e^{-2u} u^{n-2} du,$$

as long as $\frac{n-\alpha}{2} > 0$ and $n > 1$. In particular

$$\int_0^\infty e^{-2\xi_0 |z-w|^2 s} s^{\frac{n-\alpha}{2}-1} (1+s)^{\frac{n+\alpha}{2}-1} ds$$

$$\sim \xi_0^{-n+1} |z-w|^{2(n-1)}$$

if $\xi_0 > 0$, $n > 1$ and $|z-w|$ is small. Thus to have the "correct" fundamental singularity we set

$$a(\xi_0) = c_\alpha \xi_0^{n-1}. \tag{1.9}$$

We would like to point out that we are still in the process of trying to find the symbol $\sigma(G_\alpha)$ by heuristic considerations. Once found, we shall prove its correctness for all $n \geqq 1$.

We continue by applying the operator induced by the kernel

$$\hat{G}_\alpha = e^{\xi_0(z\bar{w}-\bar{z}w)} W_\alpha(\xi_0|z-w|^2)$$

to

$$(2\pi)^{-n} \int_{\mathbb{R}^{2n}} e^{i\langle y, \xi'\rangle} \hat{f}(\xi_0, \xi') d\xi', \quad \xi = (\xi_0, \xi').$$

After interchanging the order of integration, we obtain the kernel

$$(1.10) \qquad \int_{\mathbb{R}^{2n}} e^{\xi_0(z\bar{w} - \bar{z}w)} W_\alpha(\xi_0|z-w|^2)\, e^{i<y',\xi'>} dy'.$$

Now

$$\xi_0(z\bar{w} - \bar{z}w) + i <y', \xi'>$$

$$= i(2\omega_{j-1} y_{2j-1} + 2\omega_{2j} y_{2j}),$$

where $w_j = y_{2j-1} + iy_{2j}$, $j=1, \ldots, n$ and we set

$$\omega_{2j-1} = \frac{1}{2}(\xi_{2j-1} + 2x_{2j}\,\xi_0),$$

$$\omega_{2j} = \frac{1}{2}(\xi_{2j} - 2x_{2j-1}\,\xi_0),$$

$j=1, \ldots, n$. Thus (1.10) becomes

$$(1.11) \qquad \int_{\mathbb{R}^{2n}} e^{i<2\omega, y'>} W_\alpha(\xi_0|z-w|^2) dy'$$

Finally, to obtain the symbol, we multiply (1.11) by $e^{-i<x,\xi>}$, set $x_j - y_j = \sigma_j$, $j=1, \ldots, 2n$ and note that

$$<2\omega, x'> = <\xi', x'>.$$

This yields

$$\sigma(G_\alpha)$$

$$= \int_{\mathbb{R}^{2n}} e^{-i<2\omega, y'>} W_\alpha(\xi_0|y'|^2)\, dy'$$

$$= c_\alpha \int_{\mathbb{R}^{2n}} e^{-i<2\omega, y'>} dy'\, \xi_0^{n-1}$$

$$\cdot \int_0^\infty e^{-\xi_0 |y'|^2(1+2s)} s^{\frac{n-\alpha}{2}-1} (1+s)^{\frac{n+\alpha}{2}-1} ds$$

$$= c_\alpha \xi_0^{n-1} \int_0^\infty s^{\frac{n-\alpha}{2}-1} (1+s)^{\frac{n+\alpha}{2}-1} ds$$

$$\cdot \int_{\mathbb{R}^{2n}} e^{-i<2\omega, y'>} e^{-\xi_0 |y'|^2(1+2s)} dy' .$$

The interchange of the order of integration is justified by Fubini's theorem because the second iterated integral is easily seen to be convergent. Now

$$\int_{\mathbb{R}^{2n}} e^{-i<2\omega, y'>} e^{-\xi_0 |y'|^2(1+2s)} dy'$$

$$= \frac{\pi^n}{\xi_0^n (1+2s)^n} e^{-\omega^2/\xi_0(1+2s)} ,$$

which yields

$$\sigma(G_\alpha) = \frac{c_\alpha}{\xi_0} \int_0^\infty s^{\frac{n-\alpha}{2}-1} (1+s)^{\frac{n+\alpha}{2}-1} \cdot (1+2s)^{-n} e^{-\frac{|\omega|^2}{\xi_0} \frac{1}{1+2s}} ds. \tag{1.12}$$

Led by our heuristics, we state our first result as follows:

1.13. Proposition. *If* $\operatorname{Re}(\frac{n-\alpha}{2}) > 0$, *then*

(1.14) $$\sigma(G_\alpha)(x,\xi) = \frac{1}{\xi_0}\int_0^1 (1-s)^{\frac{n-\alpha}{2}-1}(1+s)^{\frac{n+\alpha}{2}-1} e^{-\frac{|\omega|^2}{\xi_0}s}\,ds$$

induces the operator

$$(G_\alpha f)(x) = (2\pi)^{-2n-1}\int_{\mathbb{R}^{2n+1}} e^{i\langle x,\xi\rangle}\sigma(G_\alpha)(x,\xi)\hat{f}(\xi)\,d\xi$$

which has the property that

$$\mathcal{L}_\alpha G_\alpha f(x) = f(x)$$

if f is in the Schwartz space and supp $\hat{f}$ is contained in $\{\xi \in \mathbb{R}^{2n+1}; \xi_0 > 0\}$

Proof. We replace $\frac{1}{1+2s}$ by s in (1.12) and obtain the form (1.14). We will show that if $c_\alpha = 1$, then G_α is a fundamental solution of $\mathcal{L}_\alpha$, i.e.,

$$\mathcal{L}_\alpha\left((2\pi)^{-2n-1}\int_{\mathbb{R}^{2n+1}} e^{i\langle x,\xi\rangle}\sigma(G_\alpha)(x,\xi)\hat{f}(\xi)\,d\xi\right) = f(x),$$

if f belongs to the Schwartz space of functions, such that supp $\hat{f} \subset \{\xi \in \mathbb{R}^{2n+1}; \xi_0 > 0\}$. This requires

$$\left\{\sum_{j=1}^n\left(|\Omega_j|^2 - \frac{\partial^2}{\partial z_j\partial\bar{z}_j}\right) - \alpha\xi_0 + \sum_{j=1}^n\left(\bar{\Omega}_j\frac{\partial}{i\partial\bar{z}_j} + \Omega_j\frac{\partial}{i\partial z_j}\right)\right\}$$

$$\circ\, \sigma(G_\alpha)(x,\xi) \equiv 1,$$

where we set

$$\Omega_j = \omega_{2j-1} + i\omega_{2j}, \quad j=1,\ldots,n.$$

To simplify matters we note that

$$\sigma(G_\alpha)(x,\xi) = \frac{c_\alpha}{\xi_0}\, g\left(\frac{|\Omega|^2}{\xi_0}\right).$$

A simple computation yields

$$\sum_{j=1}^{n}\left(\bar{\Omega}_j \frac{\partial}{i\partial \bar{z}_j} + \Omega_j \frac{\partial}{i\partial z_j}\right) h\left(|\Omega|^2\right) = 0$$

for every function h. Next we reduce the problem to solving an inhomogeneous ordinary differential equation as follows.

$$\frac{\partial^2 g\left(\frac{|\Omega|^2}{\xi_0}\right)}{\partial z_j \partial \bar{z}_j}$$

$$= g''\left(\frac{|\Omega|^2}{\xi_0}\right)\frac{1}{\xi_0^2}\,\frac{\partial(|\Omega_j|^2)}{\partial z_j}\,\frac{\partial(|\Omega_j|^2)}{\partial \bar{z}_j}$$

$$+ g'\left(\frac{|\Omega|^2}{\xi_0}\right)\frac{1}{\xi_0}\,\frac{\partial^2(|\Omega|^2)}{\partial z_j \partial \bar{z}_j}$$

$$= g''\left(\frac{|\Omega|^2}{\xi_0}\right)|\Omega_j|^2 + g'\left(\frac{|\Omega|^2}{\xi_0}\right)\xi_0 .$$

Therefore

$$e^{-i\langle x,\xi\rangle}\mathcal{L}_\alpha\left(e^{i\langle x,\xi\rangle}\sigma(G_\alpha)(x,\xi)\right)$$

$$= c_\alpha\left(\frac{|\Omega|^2}{\xi_0}\, g\left(\frac{|\Omega|^2}{\xi_0}\right) - \frac{|\Omega|^2}{\xi_0}\, g''\left(\frac{|\Omega|^2}{\xi_0}\right)\right.$$

$$\left. - ng'\left(\frac{|\Omega|^2}{\xi_0}\right) - \alpha g\left(\frac{|\Omega|^2}{\xi_0}\right)\right).$$

We set $|\Omega|^2/\xi_0 = x$ and require that the following differential equation is satisfied

(1.15) $$c_\alpha(xg''(x) + ng'(x) + (\alpha - x)\, g(x)) = -1.$$

Following previous calculations we set

$$g(x) = \int_0^1 (1-s)^{\frac{n-\alpha}{2} - 1} (1+s)^{\frac{n+\alpha}{2} - 1} e^{-xs}\, ds.$$

Then the left-hand side of (1.15) becomes

$$c_\alpha \int_0^1 (1-s)^{\frac{n-\alpha}{2} - 1} (1+s)^{\frac{n+\alpha}{2} - 1} e^{-xs} \cdot (xs^2 - ns + \alpha - x)\, ds$$

$$= c_\alpha \int_0^1 \frac{d}{ds}\left((1-s)^{\frac{n-\alpha}{2} - 1} (1+s)^{\frac{n+\alpha}{2} - 1} e^{-xs}\right) ds$$

$$= -c_\alpha ,$$

which yields $c_\alpha = 1$. This proves Proposition 1.13.

Next we continue $\sigma(G_\alpha)(x, \xi)$ analytically on the complex α-plane. Let D denote the contour

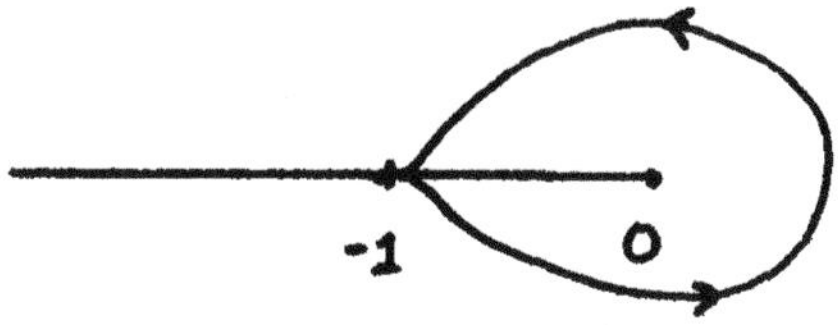

In other words D starts at -1, encircles the origin once counterclockwise and returns to -1. Consider

(1.16) $$\psi_\alpha(x) = \frac{1}{\xi_0}\int_D s^{\frac{n-\alpha}{2}-1}(2+s)^{\frac{n+\alpha}{2}-1} e^{-x(1+s)}\,ds.$$

Again we apply $\mathcal{L}_\alpha$ to ψ_α as we did in (1.15), we obtain

(1.17) $$-1 = c_\alpha(x\psi_\alpha''(x) + n\psi_\alpha'(x) + (\alpha-x)\psi_\alpha(x))$$

$$= c_\alpha\int_D s^{\frac{n-\alpha}{2}-1}(2+s)^{\frac{n+\alpha}{2}-1} e^{-x(1+s)}$$

$$\cdot\,(x(1+s)^2 - n(1+s) + \alpha - x)\,ds$$

$$= -c_\alpha\int_D \frac{d}{ds}\left(s^{\frac{n-\alpha}{2}}(2+s)^{\frac{n+\alpha}{2}} e^{-x(1+s)}\right)ds$$

$$= -c_\alpha\left[s^{\frac{n-\alpha}{2}}(2+s)^{\frac{n+\alpha}{2}} e^{-x(1+s)}\right]_{D_{\mathrm{in.}}}^{D_{\mathrm{end.}}}$$

On D we set

$$s^a = e^{a\log s}$$

where $\log s$ is the principal branch of the logarithm, i.e., $\log s$ is real if s is on the positive part of the real axis Therefore

$$(1.17) = -c_\alpha\left(e^{\frac{n-\alpha}{2}(\log 1+\pi i)} - e^{\frac{n-\alpha}{2}(\log 1-\pi i)}\right)$$

$$= -c_\alpha\cdot 2i\sin\left(\frac{n-\alpha}{2}\pi\right).$$

<u>1.18. Proposition.</u> <u>If</u> $\mathrm{Re}\,(n-\alpha) > 0$, $\frac{n-\alpha}{2} \neq 0, \pm 1, \pm 2, \ldots$ <u>then for</u> $\xi_0 > 0$

(1.19) $$\sigma(G_\alpha)(x, \xi)$$

$$= \frac{1}{2i} \operatorname{cosec}\left(\frac{n-\alpha}{2}\pi\right) \xi_0^{-1} \int_D s^{\frac{n-\alpha}{2}-1} (2+s)^{\frac{n+\alpha}{2}-1} e^{-\frac{|\Omega|^2}{\xi_0}(1+s)} ds.$$

$\sigma(G_\alpha)(x,\xi)$ can be continued analytically on the complex α-plane to all α, such that

$$\frac{n-\alpha}{2} \neq 0, -1, -2, \ldots$$

Proof. In (1.19) we deform the path of integration, D, into integrating along the real axis from -1 to $-\delta$, $0 < \delta < 1$, then integrating on the circle of radius δ around the origin and, finally returning from $-\delta$ to -1. On the first part

$$s^{\frac{n-\alpha}{2}-1} = e^{-\pi i(\frac{n-\alpha}{2}-1)} (-s)^{\frac{n-\alpha}{2}-1},$$

where $\log(-s)$ is real, on the last part

$$s^{\frac{n-\alpha}{2}-1} = e^{\pi i(\frac{n-\alpha}{2}-1)} (-s)^{\frac{n-\alpha}{2}-1}.$$

Therefore

$$\int_D s^{\frac{n-\alpha}{2}-1} (2+s)^{\frac{n+\alpha}{2}-1} e^{-x(1+s)} ds$$

$$= e^{-i\pi(\frac{n-\alpha}{2}-1)} \int_{-1}^{-\delta} (-s)^{\frac{n-\alpha}{2}-1} (2+s)^{\frac{n+\alpha}{2}-1} e^{-x(1+s)} ds$$

$$+ \int_{-\pi}^{\pi} (\delta e^{i\theta})^{\frac{n-\alpha}{2}-1} (2+\delta e^{i\theta})^{\frac{n+\alpha}{2}-1} e^{-x(1+\delta e^{i\theta})} d(\delta e^{i\theta})$$

$$+ e^{i\pi(\frac{n-\alpha}{2}-1)} \int_{-\delta}^{-1} (-s)^{\frac{n-\alpha}{2}-1} (2+s)^{\frac{n+\alpha}{2}-1} e^{-x(1+s)} ds$$

If $\operatorname{Re}\frac{n-\alpha}{2} > 0$ we let $\delta \longrightarrow 0$ and the contribution of the integral on the circle vanishes. Thus we are left with

$$(1.20) \qquad -2i\sin\pi\left(\frac{n-\alpha}{2}-1\right)$$

$$\cdot \int_{-1}^{0} (-s)^{\frac{n-\alpha}{2}-1} (2+s)^{\frac{n+\alpha}{2}-1} e^{-x(1+s)} ds .$$

We set $-t = 1-s$ and (1.20) combined with (1.19) yields (1.14). This proves the proposition.

Analogous calculations yield the symbol $\sigma(G_\alpha)(x,\xi)$ for $\xi_0 < 0$, as we have already remarked.

We collect the results of this chapter in the following form.

__1.21 Theorem.__ _The symbol_ $\sigma(G_\alpha)(x,\xi)$ _of a fundamental solution of_ $\mathcal{L}_\alpha$ _is given by_

$$(1.22) \qquad \sigma(G_\alpha)(x,\xi) = \hat{k}_\alpha(\xi_0, 2\omega_1, \ldots, 2\omega_{2n}),$$

$$(1.23) \qquad \xi_0 \neq 0,$$

where

$$\omega_{2j-1} = \frac{1}{2}\xi_{2j-1} + x_{2j}\xi_0,$$

$$\omega_{2j} = \frac{1}{2}\xi_{2j} - x_{2j-1}\xi_0$$

for $j=1,\ldots,n,$ _and_ $\hat{k}_\alpha$ _is given by_

$$(1.24) \qquad \hat{k}_\alpha(\sigma) = \frac{1}{2i\sin\left(\frac{n-\alpha\operatorname{sign}\sigma_0}{2}\pi\right)}$$

$$\cdot \, |\sigma_0|^{-1} \int_D s^{\frac{n-\alpha \operatorname{sign} \sigma_0}{2} - 1} (2+s)^{\frac{n+\alpha \operatorname{sign} \sigma_0}{2} - 1} e^{-\frac{|\sigma'|^2}{4|\sigma_0|}(1+s)} \, ds$$

<u>if</u> $\frac{n-\alpha \operatorname{sign} \sigma_0}{2} \neq 0, \pm 1, \pm 2, \ldots$. <u>Here</u> $\sigma = (\sigma_0, \sigma')$. <u>Moreover,</u> $\hat{k}_\alpha(\sigma)$ <u>can be continued analytically in the complex</u> α <u>plane to</u> $\frac{n-\alpha \operatorname{sign} \sigma_0}{2} = 1, 2, 3, \ldots$ <u>according to the formula</u>

(1.25)

$$\hat{k}_\alpha(\sigma) = \frac{1}{|\sigma_0|} \int_0^1 (1-s)^{\frac{n-\alpha \operatorname{sign} \sigma_0}{2} - 1} (1+s)^{\frac{n+\alpha \operatorname{sign} \sigma_0}{2} - 1} \cdot e^{-\frac{|\sigma'|^2}{4|\sigma_0|} s} \, ds$$

<u>which is also valid for all</u> α <u>with</u> $n > \operatorname{Re} \alpha \operatorname{sign} \sigma_0$.

Observe that when $\sigma_0 \ (=\xi_0) \neq 0$, then k_α is defined by either (1.24) or (1.25) as long as $\frac{n \pm \alpha}{2} \neq 0, -1, -2, \ldots$.

<u>1.26 Corollary.</u> $\hat{k}_\alpha(\sigma_0, \sigma')$ <u>can be extended by continuity to</u> $\sigma_0 = 0$ <u>as follows</u>

(1.27) $$\hat{k}_\alpha(0, \sigma') = \frac{4}{|\sigma'|^2}.$$

<u>Thus defined</u> $\hat{k}_\alpha(\sigma)$ <u>is</u> C^∞ <u>outside of the origin.</u>

<u>Proof.</u> This follows from a simple integration by parts. For example, if $\operatorname{Re} \frac{n-\alpha}{2} > 1$, then

$$\text{(1.28)} \qquad \hat{k}_\alpha(\sigma) = \frac{4}{|\sigma'|^2}\left(1 + \int_0^1 \frac{d}{ds}\left((1-s)^{\frac{n-\alpha}{2}-1}(1+s)^{\frac{n+\alpha}{2}-1}\right) e^{-\frac{|\sigma'|^2}{4\sigma_0}s}\, ds\right)$$

This proves Corollary 1.26.

Chapter 2. A comparison

Set

$$(2.1)\qquad \Phi_\alpha = \frac{1}{c_\alpha}\varphi_\alpha = \frac{\Gamma(\frac{n+\alpha}{2})\,\Gamma(\frac{n-\alpha}{2})}{2^{2-2n}\pi^{n+1}}\varphi_\alpha ,$$

where

$$(2.2)\qquad \varphi_\alpha = \frac{1}{(|z|^2 - it)^{\frac{n+\alpha}{2}}(|z|^2 + it)^{\frac{n-\alpha}{2}}}$$

In (2.1) we assume $\frac{n+\alpha}{2} \neq 0, -1, -2, \dots$. According to Proposition 7.1 of [9] the operator K_α, defined by

$$(2.3)\qquad K_\alpha f = f * \Phi_\alpha = c_\alpha^{-1}\int_{H_n} f(y)\varphi_\alpha(y^{-1}x)\,dH_n(y),$$

$f \in C_0^\infty(H_n)$ is inverse to $\mathcal{L}_\alpha$, as long as α is admissible, i.e., $\frac{n\pm\alpha}{2} \neq 0, -1, -2, \dots$.

2.4 Theorem. *Let α be admissible. Then*

$$\sigma(K_\alpha)(x,\xi) = \sigma(G_\alpha)(x,\xi),$$

$\xi \neq 0$, *where* $\sigma(G_\alpha)(x,\xi)$ *is given by* (1.22). *In other words,*

$$(2.5)\qquad (K_\alpha f)(x) = (2\pi)^{-2n-1}\int_{\mathbb{R}^{2n+1}} e^{i\langle x,\xi\rangle}\sigma(G_\alpha)(x,\xi)\hat{f}(\xi)\,d\xi$$

whenever $f \in \mathcal{S}$,

Proof. Starting with

$$(2.6)\qquad (K_\alpha f)(x) = c_\alpha^{-1}(2\pi)^{-2n-1}\int_{H_n} \varphi_\alpha(y^{-1}x)\,dH_n(y)\int_{\mathbb{R}^{2n+1}} e^{i<y,\xi>}\hat{f}(\xi)\,d\xi$$

we shall compute

$$(2.7)\qquad \sigma(K_\alpha)(x,\xi) = c_\alpha^{-1}e^{-i<x,\xi>}\int_{\mathbb{R}^{2n+1}} \varphi_\alpha(y^{-1}x)e^{i<y,\xi>}dy\,2^{-n}.$$

We shall do the computation only if $\xi_0 > 0$. It is similar if $\xi_0 < 0$. We use the notation

$$z_j = x_{2j-1} + ix_{2j}\,,\quad w_j = y_{2j-1} + iy_{2j},\quad j=1,\dots,n,$$

and recall that

$$dH_n(x) = 2^{-n}dx.$$

First we compute

$$(2.8)\qquad c_\alpha^{-1}\int_{-\infty}^{\infty}\varphi_\alpha(y^{-1}x)\,e^{i(y_0-x_0)\xi_0}dy_0\cdot 2^{-n}$$

$$= c_\alpha^{-1}e^{\xi_0(z\bar{w}-\bar{z}w)}\cdot 2^{-n}$$

$$\cdot\int_{-\infty}^{\infty} e^{it\xi_0}(|z-w|^2+it)^{-\frac{n+\alpha}{2}}(|z-w|^2-it)^{-\frac{n-\alpha}{2}}dt$$

$$= c_\alpha^{-1}e^{\xi_0(z\bar{w}-\bar{z}w)}2^{-n}|z-w|^{2(1-n)}$$

$$\cdot\int_{-\infty}^{\infty} e^{is(\xi_0|z-w|^2)}(1+is)^{-\frac{n+\alpha}{2}}(1-is)^{-\frac{n-\alpha}{2}}ds.$$

Hence we need to evaluate

$$(2.9)\qquad \int_{-\infty}^{\infty} e^{isx}(1+is)^{-\frac{n+\alpha}{2}}(1-is)^{-\frac{n-\alpha}{2}}ds,\quad x>0.$$

Now

$$(2.9) = i\int_{-i\infty}^{i\infty} e^{-xz}(1+z)^{-\frac{n-\alpha}{2}}(1-z)^{-\frac{n+\alpha}{2}}dz$$

$$= -2^{1-n}i\int_{-i\infty}^{i\infty} e^{-2xu}\left(\frac{1}{2}+u\right)^{-\frac{n-\alpha}{2}}\left(\frac{1}{2}-u\right)^{-\frac{n+\alpha}{2}}du,$$

where we set $z = 2u$. If $-\frac{n+\alpha}{2} > 0$ we can change the path of integration from $(-i\infty, i\infty)$ to $(\frac{1}{2} - i\infty, \frac{1}{2}+i\infty)$, so

$$(2.9) = -2^{1-n}i\int_{\frac{1}{2}-i\infty}^{\frac{1}{2}+i\infty} e^{-2xu}\left(\frac{1}{2}+u\right)^{-\frac{n-\alpha}{2}}\left(\frac{1}{2}-u\right)^{-\frac{n+\alpha}{2}}du$$

$$= -2^{1-n}ie^{-x}\int_{-i\infty}^{i\infty} e^{2xv}v^{-\frac{n+\alpha}{2}}(1-v)^{-\frac{n-\alpha}{2}}dv$$

$$= -2^{1-n}ie^{-x}\int_{-i\infty}^{i\infty} e^{-2xv}(-v)^{-\frac{n+\alpha}{2}}(1+v)^{-\frac{n-\alpha}{2}}dv$$

$$= -2^{1-n}ie^{-x}\int_{L} e^{-2xv}(-v)^{-\frac{n+\alpha}{2}}(1+v)^{-\frac{n-\alpha}{2}}dv,$$

where L denotes the following path

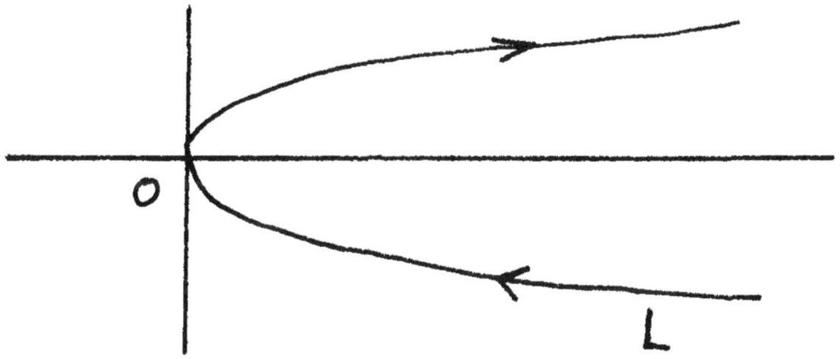

This deformation is justified by noting that for $|v| = R$, R large, the integrand is bounded by

$$\frac{e^{-2Rx\cos\theta}}{R}, \quad 0 \leqq \theta \leqq \frac{\pi}{2},$$

since $n \geqq 1$. Finally $(-v)^{\gamma} = e^{\gamma \log(-v)}$, where log denotes the principal branch of the logarithm, i.e., $\log(-v) > 0$ if $v < 0$. This easily yields

$$(2.9) = \left(e^{\pi i \frac{n+\alpha}{2}} - e^{-\pi i \frac{n+\alpha}{2}}\right)(-1)\,2^{1-n}$$

$$\cdot\, ie^{-x} \int_0^{\infty} e^{-2xv} v^{-\frac{n+\alpha}{2}} (1+v)^{-\frac{n-\alpha}{2}} dv.$$

Thus, we have derived

<u>2.10 Lemma.</u> <u>Let</u> $x > 0$ <u>and</u> $\frac{n+\alpha}{2} < 0$. <u>Then</u>

$$\int_{-\infty}^{\infty} e^{isx} (1+is)^{-\frac{n+\alpha}{2}} (1-is)^{-\frac{n-\alpha}{2}} ds$$

$$= 2^{2-n} e^{-x} \sin\left(\frac{n+\alpha}{2}\pi\right)$$

$$\cdot \int_0^{\infty} e^{-2xs} s^{-\frac{n+\alpha}{2}} (1+s)^{-\frac{n-\alpha}{2}} ds.$$

The next result concerns changing the parameters n and α.

<u>2.11 Lemma.</u> <u>Let</u> a, c <u>be real numbers,</u> $a > 0$, <u>and</u> $a - c > 0$. <u>Suppose</u> $x > 0$. <u>Then</u>

$$(2.12) \qquad \frac{1}{\Gamma(a-c+1)} \int_0^{\infty} e^{-xt} t^{a-c} (1+t)^{-a} dt$$

$$= x^{c-1} \frac{1}{\Gamma(a)} \int_0^{\infty} e^{-xt} t^{a-1} (1+t)^{c-a-1} dt.$$

<u>Proof.</u> Erdelyi: [6], v. 1. p. 256 proves the following formula

$$(2.13) \qquad \frac{1}{\Gamma(a)} \int_0^{\infty} e^{-xt} t^{a-1} (1+t)^{c-a-1} dt$$

$$= \frac{1}{2\pi i} \int_{\gamma-i\infty}^{\gamma+i\infty} \frac{\Gamma(a+s)\,\Gamma(-s)\,\Gamma(1-c-s)}{\Gamma(a)\,\Gamma(a-c+1)} x^s ds$$

where $-a < \gamma < \min(0,-c)$. We replace a by $a-c+1$, and c by $c-2$ in (2.13), we obtain

$$(2.14) \qquad \frac{1}{\Gamma(a-c+1)} \int_0^{\infty} e^{-xt} t^{a-c} (1+t)^{-a} dt$$

$$= \frac{1}{2\pi i} \int_{\omega-i\infty}^{\omega+i\infty} \frac{\Gamma(a-c+1+s)\,\Gamma(-s)\,\Gamma(-1+c-s)}{\Gamma(a-c+1)\,\Gamma(a)} x^s ds$$

as long as $-a+c-1 < \omega < \min(0, c-2)$. Actually, since $a > 0$ and $a > c$ we may assume that ω satisfies the more restrictive condition

$$-a+c-1 < \omega < \min(-1, c-1).$$

Setting $t=s+1-c$ in the right-hand side of (2.14) we obtain that the left-hand side of (2.14) is equal to

$$\frac{1}{2\pi i} \int_{\omega+1-c-i\infty}^{\omega+1-c+i\infty} \frac{\Gamma(a+t)\,\Gamma(1-c-t)\,\Gamma(-t)\, x^{t+c-1}}{\Gamma(a+c-1)\,\Gamma(a)} dt$$

which gives the right-hand side of (2.12), since $-a < \gamma = \omega+1-c < \min(-c, 0)$. This proves Lemma 2.11.

Next we return to the computation of $\sigma(K_\alpha)(x,\xi)$. Set $a = 1 - \frac{n+\alpha}{2}$ and $c = 2-n$. Then the hypotheses of Lemma 2.11 can be put in the form

$$1 - \frac{n+\alpha}{2} > 0, \quad \frac{n-\alpha}{2} - 1 > 0,$$

or, equivalently,

(2.15) $$\alpha < \min(-1, 2-n).$$

According to Lemma 2.11 we have

$$\int_0^\infty e^{-2xs} s^{-\frac{n+\alpha}{2}} (1+s)^{-\frac{n-\alpha}{2}} ds$$

$$= \frac{\Gamma(1 - \frac{n+\alpha}{2})}{\Gamma(\frac{n-\alpha}{2})} (2x)^{-1+n} \int_0^\infty e^{-2xs} s^{\frac{n-\alpha}{2}-1} (1+s)^{\frac{n+\alpha}{2}-1} ds.$$

Thus Lemma 2.10 yields

<u>2.16 Lemma.</u> <u>Let</u> $\alpha < \min(-1, 2-n)$. <u>Then</u>

(2.17) $$\int_{-\infty}^\infty e^{isx} (1+is)^{-\frac{n+\alpha}{2}} (1-is)^{-\frac{n-\alpha}{2}} ds$$

$$= \frac{2\pi e^{-x} x^{n-1}}{\Gamma(\frac{n+\alpha}{2}) \Gamma(\frac{n-\alpha}{2})}$$

$$\cdot \int_0^\infty e^{-2xs} s^{\frac{n-\alpha}{2}-1} (1+s)^{\frac{n+\alpha}{2}-1} ds.$$

Now we are ready to complete the proof of Theorem 2.4. First (2.7) and (2.17) yield

(2.18) $$\sigma(K_\alpha)(x,\xi)$$

$$= \frac{2^{n-1}\xi_0^{n-1}}{\pi^n} \int_{\mathbb{R}^{2n}} e^{i\langle y' - x', \xi'\rangle} e^{-\xi_0|z-w|^2 + \xi_0(z\bar{w} - \bar{z}w)} dy'$$

$$\cdot \int_0^\infty e^{-2\xi_0|z-w|^2 s} s^{\frac{n-\alpha}{2}-1} (1+s)^{\frac{n+\alpha}{2}-1} ds.$$

Next the analysis that yields (1.14) from (1.11) applies and we obtain

(2.19) $$\sigma(K_\alpha)(x,\xi) = \frac{1}{\xi_0} \int_0^1 (1-s)^{\frac{n-\alpha}{2}-1} (1+s)^{\frac{n+\alpha}{2}-1} e^{-\frac{|\Omega|^2}{\xi_0}s} ds,$$

as long as $\xi_0 > 0$ and

$$\alpha < \min(-1, 2-n).$$

To remove the restriction on α, we note that

$$\varphi_\alpha(z,t) = (|z|^4 + t^2)^{-n/2} \left(\frac{|z|^2 + it}{|z|^2 - it}\right)^{\alpha/2}$$

and by introducing parabolic coordinates with $v = (|z|^4 + t^2)^{1/4}$, φ_α is integrable at the origin and entire in α. Therefore

(2.20) $$K_\alpha f(x) = c_\alpha^{-1} \int_{H_n} f(xu^{-1}) \varphi_\alpha(u)\, dH_n(u)$$

is holomorphic in α whenever α is admissible.

Next we note that $\sigma(K_\alpha)(x,\xi)$ given by (2.19) if $\xi_0 > 0$ and $\alpha < \min(-1, 2-n)$ can be extended to all admissible α holomorphically, the extension given in Theorem 1.22, also for $\xi_0 < 0$. From the formula (1.23) it follows immediately that if $f(x)$ satisfies the hypothesis of

Theorem 2.4 then $(K_\alpha f)(x)$ given by

$$(2.21) \qquad (2\pi)^{-2n-1} \int_{\mathbb{R}^{2n+1}} e^{i<x,\xi,>} \sigma(G_\alpha)(x,\xi)\hat{f}(\xi)\,d\xi$$

is holomorphic in α when α is admissible. Since (2.20) and (2.21) agree when α is in an open interval of the real axis, they must agree for all admissible α. This proves Theorem 2.4.

2.22 Corollary. Set $x = (x_0, x') = (x_0, x_1, \ldots, x_{2n})$ and $z_j = x_{2j-1} + ix_{2j}$, $j=1,\ldots,n$. Then

$$\hat{k}_\alpha(\sigma_0, \sigma_1, \ldots, \sigma_{2n}) = \frac{\Gamma(\frac{n+\alpha}{2})\Gamma(\frac{n-\alpha}{2})}{2^{2-2n}\pi^{n+1}}$$

$$\cdot \int_{\mathbb{R}^{2n+1}} e^{-i<x,\sigma>} (|z|^2 - ix_0)^{-\frac{n+\alpha}{2}} (|z|^2 + ix_0)^{-\frac{n-\alpha}{2}} 2^{-n} dx$$

where $\hat{k}_\alpha(\sigma)$ is defined in (1.24).

As a by-product of these considerations we found a solution of an inhomogeneous confluent hypergeometric differential equation. We recall that in Whittaker's standard form the confluent hypergeometric differential equation is given by

$$(2.23) \qquad \frac{d^2y}{dx^2} + \left(-\frac{1}{4} + \frac{k}{x} + \frac{1/4 - \mu^2}{x^2}\right) y = 0.$$

Let us consider the differential equation (1.15)

$$xh''(x) + nh'(x) + (\alpha - x)h(x) = -1.$$

We substitute

$$h(x) = x^{-n/2} y(2x).$$

Then an elementary calculation yields the following result.

2.24 Proposition. *Let* $x > 0$. *Then*

$$(2.25) \qquad y_{n,\alpha}(x) = \left(\frac{x}{2}\right)^{n/2} \int_0^1 (1-s)^{\frac{n-\alpha}{2}-1} (1+s)^{\frac{n+\alpha}{2}-1} e^{-\frac{x}{2}s} ds$$

for $\operatorname{Re}\frac{n-\alpha}{2} > 0$ *and its analytic continuation for all other* α, $\alpha \neq n, n+2,$ $n+4, \ldots$ *yields a solution of the following inhomogeneous Whittaker differential equation*

$$(2.26) \qquad \frac{d^2y}{dx^2} + \left(-\frac{1}{4} + \frac{\frac{1}{2}\alpha}{x} + \frac{\frac{1}{4} - \left(\frac{n-1}{2}\right)^2}{x^2}\right) y = -\left(\frac{x}{2}\right)^{n/2+1}.$$

The analytic continuation is given by

$$(2.27) \qquad y_{n,\alpha}(x) = \frac{1}{2i\sin\left(\frac{n-\alpha}{2}\pi\right)} \cdot \int_D t^{\frac{n-\alpha}{2}-1} (2+t)^{\frac{n+\alpha}{2}-1} e^{-\frac{x}{2}(1+t)} dt$$

if $\frac{n-\alpha}{2} \neq 0, \pm 1, \pm 2, \ldots,$ *where* D *is the Hankel contour appearing in* (1.16).

Chapter 3. $\Box_b$ on functions and the solvability of the Lewy equation

A q-form f on H_n is a sum

$$(3.1) \qquad f = \sum_{|J|=q} f_J \, d\bar{z}^J$$

where f_J are complex-valued functions on H_n indexed by $J = (j_1, j_2, \ldots, j_q)$, $j_1 < j_2 < \ldots < j_q$, and $d\bar{z}^J = d\bar{z}_{j_1} \wedge d\bar{z}_{j_2} \wedge \ldots \wedge d\bar{z}_{j_q}$.

The $\bar{\partial}_b$ operator (mapping q-forms to q+1-forms) is then defined by

$$(3.2) \qquad \bar{\partial}_b f = \sum_{j,J} \bar{Z}_j(f_J)\, d\bar{z}_j \wedge d\bar{z}^J,$$

where $Z_j = \frac{\partial}{\partial z_j} + i\bar{Z}_j \frac{\partial}{\partial t}$, $j=1,\ldots,n$, see (1.1), and the formal adjoint ϑ_b is given by

$$(3.3) \qquad \vartheta_b f = -\sum_{j,J} Z_j(f_J)\, d\bar{z}_j \,\lrcorner\, d\bar{z}^J$$

The interior product $\lrcorner$ is defined after formula (6.10). One then defines the Laplacian corresponding to this complex; it is

$$(3.4) \qquad \Box_b = \bar{\partial}_b \vartheta_b + \vartheta_b \bar{\partial}_b .$$

We denote the restriction of $\Box_b$ to q-forms by $\Box_b^{(q)}$. It turns out that on the Heisenberg group $\Box_b^{(q)}$ takes a particularly elegant form. $\Box_b^{(q)}$ is diagonal, more precisely

$$(3.5) \qquad \Box_b^{(q)}\Big(\sum_J f_J \, d\bar{z}^J\Big) = \sum_J \mathcal{L}_\alpha(f_J)\, d\bar{z}^J$$

with $\alpha = n-2q$, where $\mathcal{L}_\alpha$ is defined by (1.2). The regularity and the existence theory for $\mathcal{L}_\alpha$ has been studied in [9], using the fundamental solution discussed in Chapter 2. *This yields a fundamental solution for* $\square_b^{(q)}$ if $0 < q < n$. Such a fundamental solution does not exist for $q=0$ or $q=n$--these correspond to $\alpha = \pm n$ for $\mathcal{L}_\alpha$. We shall now consider these cases. Actually we shall do the work only for $q=0$, because $q=n$ follows by complex conjugation.

A simple calculation shows that

$$\square_b^{(0)} = -\sum_{j=1}^{n} Z_j \overline{Z}_j. \tag{3.6}$$

$\square_b^{(0)}$ therefore annihilates the boundary values of holomorphic functions on H_n, and so is far from hypoelliptic. Moreover (we shall see below) the equation

$$\square_b^{(0)}(u) = f \tag{3.7}$$

is generally not even locally solvable. To get to the solution of this problem let us recall some of the basic terminology. The generalized upper half-plane $\mathcal{D} \subset \mathbb{C}^{n+1}$ is given by

$$\mathcal{D} = \{(z_1, \ldots, z_{n+1});\ \operatorname{Im} z_{n+1} > |z_1|^2 + \ldots + |z_n|^2\} \tag{3.7'}$$

with its boundary $b\mathcal{D}$,

$$b\mathcal{D} = \{(z_1, \ldots, z_{n+1});\ \operatorname{Im} z_{n+1} = |z_1|^2 + \ldots + |z_n|^2\}. \tag{3.8}$$

The mapping $(z,t) \in \mathbb{C}^n \times \mathbb{R} \longrightarrow (z_1', \ldots, z_{n+1}') \in \mathbb{C}^{n+1}$, given by $(z_1', \ldots, z_n') = (z_1, \ldots, z_n)$, $z_{n+1} = t + i \sum_{j=1}^{n} |z_j|^2$, identifies $\mathbb{C}^n \times \mathbb{R}$ with $b\mathcal{D}$. $\mathbb{C}^n \times \mathbb{R}$ has the structure of a group, the Heisenberg group H_n; in particular the left-invariant vectorfields $Z_j = \frac{\partial}{\partial z_j} + i\bar{z}_j \frac{\partial}{\partial t}$, $j=1,\ldots,n$, on the Heisenberg group are restrictions to $b\mathcal{D}$ of the holomorphic vectorfields $\frac{\partial}{\partial z_j} + 2i\bar{z}_j \frac{\partial}{\partial z_{n+1}}$, $j=1,\ldots,n$, and the latter are tangential at $b\mathcal{D}$.

Any $f \in L^2(H_n)$ (or a distribution with compact support) leads to the Cauchy-Szegö integral $C(f)$, defined in $\mathcal{D}$ by

$$C(f)(z) = \int_{b\mathcal{D}} S(z,w) f(w)\, d\sigma_w, \tag{3.9}$$

with

$$S(z,w) = c\left[i(\bar{w}_{n+1} - z_{n+1}) - 2\sum_{k=1}^{n} z_k \bar{w}_k\right]^{-n-1}, \tag{3.10}$$

where $c = 2^{n-1}\Gamma(n+1)/\pi^{n+1}$, f is defined on $b\mathcal{D}$ via the identification of the latter with H_n, and $d\sigma_w$ corresponds to the Lebesgue volume of H_n. The restriction of $C(f)$ to $b\mathcal{D}$ (taken on the Heisenberg group) is given by

$$C_b(f) = \lim_{\epsilon\to 0} f * S_\epsilon, \tag{3.11}$$

where $S_\epsilon = c(\epsilon^2 + |z|^2 - it)^{-n-1}$, and the convolution in (3.11) is with respect to the Heisenberg group. When $f \in L^2$, the limit in (3.11) exists in L^2-norm (for further details see Koranyi and Vagi [22]).

Next we shall derive a "relative fundamental solution" of $\Box_b^{(0)}$. We write

$$\Box_b^{(0)} = \mathcal{L}_\alpha - (\alpha - n)iT, \quad T = \frac{\partial}{\partial t}.$$

Therefore

(3.12) $$\Box_b^{(0)}(\varphi_\alpha) = \mathcal{L}_\alpha(\varphi_\alpha) - (\alpha-n)\, iT(\varphi_\alpha)$$
$$= c_\alpha \delta - (\alpha-n)\, iT(\varphi_\alpha),$$

where

(3.13) $$\varphi_\alpha = (|z|^2 - it)^{-\frac{n+\alpha}{2}} (|z|^2 + it)^{-\frac{n-\alpha}{2}},$$

and

(3.14) $$c_\alpha = \frac{2^{2-n}\pi^{n+1}}{\Gamma(\frac{n+\alpha}{2})\Gamma(\frac{n-\alpha}{2})}.$$

These formulas can be found in Chapter 2. Formal differentiation of (3.12) with respect to α yields the following equation

(3.15) $$\Box_b^{(0)}\left(2^{n-2}\frac{\Gamma(n)}{\pi^{n+1}}\log\left(\frac{|z|^2-it}{|z|^2+it}\right) \times (|z|^2 - it)^{-n}\right)$$
$$= \delta - c(|z|^2 - it)^{-n-1}.^{*}$$

Set

(3.16) $$\Phi = c_0 \log\left(\frac{|z|^2 - it}{|z|^2 + it}\right) \times (|z|^2 - it)^{-n}$$

where we set $c_0 = 2^{n-2}\Gamma(n)/\pi^{n+1}$.

Define the operator K by

(3.17) $$K(f) = f * \Phi.$$

<u>3.18 Lemma</u>

(3.19) $$\Box_b^{(0)} \cdot K = K \cdot \Box_b^{(0)} = I - C_b$$

<u>when acting on distributions of compact support.</u>

---

*Here the logarithm of the quotient means the difference of the corresponding logarithms.

Proof. The argument is similar to the proof of Theorem 6.2 of [9] We shall only show that $\square_b^{(0)} \cdot K = I - C_b$. The other identity follows by transposition. We set $\rho_\varepsilon = \varepsilon^2 + |z|^2 - it$. We shall then take the limit in the sense of distributions, as $\varepsilon \to 0$. Now

$$-\sum_{j=1}^{n} Z_j \overline{Z}_j \left(c_0 \frac{\log \rho_\varepsilon}{\rho_\varepsilon^n}\right) = 0.$$

On the other hand

$$Z_j \overline{Z}_j \frac{\log \overline{\rho}_\varepsilon}{\rho_\varepsilon^n} = \frac{-4n|z_j|^2 + 2|z|^2 + 2\varepsilon^2 - 2it}{\rho_\varepsilon^{n+1} \overline{\rho}_\varepsilon}.$$

Therefore

$$\sum_{j=1}^{n} Z_j \overline{Z}_j \left(\frac{\log \overline{\rho}_\varepsilon}{\rho_\varepsilon^n}\right) = 2n \left(\frac{2\varepsilon^2}{\rho_\varepsilon^{n+1} \overline{\rho}_\varepsilon} - \frac{1}{\rho_\varepsilon^{n+1}}\right).$$

Letting $\varepsilon \to 0$ we obtain

$$-\sum_{j=1}^{n} Z_j \overline{Z}_j \left[c_0 (\log \frac{\rho}{\overline{\rho}}) \rho^{-n}\right]$$

$$= 4n c_0 \left(\int_{\mathbb{R}^{2n+1}} (|z|^2 + 1 - it)^{-n-1} (|z|^2 + 1 + it)^{-1} d\sigma_w\right)\delta$$

$$- 2nc_0 \frac{1}{\rho^{n+1}}.$$

According to pp. 441-443 of [9]

$$\int_{\mathbb{R}^{2n+1}} (|z|^2 + 1 - it)^{-n-1} (|z|^2 + 1 + it)^{-1} d\sigma_w$$

$$= \int_{\mathbb{R}^{2n}} (|z|^2 + 1)^{-n-1} dxdy \int_{-\infty}^{\infty} (1 - it)^{-n-1} (1 + it)^{-1} dt$$

$$= 2^{-n}\pi^{n+1}/\Gamma(n+1),$$

where we set $z_j = x_j + iy_j$, $j=1,\ldots,n$. Hence

$$\Box_b^{(0)}\left(c_0\left[\log\frac{\rho}{\bar{\rho}}\right]\rho^{-n}\right)$$

$$= \delta - 2^{n-1}(\Gamma(n+1)/\pi^{n+1})\,\rho^{-n-1},$$

which proves Lemma 3.18.

In addition to the lemma one should observe that $C_b^2 = C_b$, $C_b^* = C_b$, $\Box_b^{(0)} C_b = C_b \Box_b^{(0)} = 0$. Thus C_b is an orthogonal projection which commutes with $\Box_b^{(0)}$. The meaning of Lemma 3.18 is that K inverts $\Box_b^{(0)}$ on the subspace orthogonal to the boundary values of holomorphic functions.

3.20 Theorem. $\Box_b^{(0)}(u) = f$ is solvable in a neighborhood of a $P \in H_n$ if and only if $C_b(f)$ is real-analytic in a neighborhood of P.

3.21 Corollary. If the above condition for f is satisfied and if f belongs to one of the spaces (see ref. [9] for definitions) $S_k^p(\Omega\text{loc})$, $\Gamma_\alpha(\Omega\text{loc})$, $L^\infty(\Omega\text{loc})$, or $C^\infty(\Omega)$, then we can find a u which belongs to $S_{k+2}^p(\Omega\text{loc})$, $\Gamma_{\alpha+2}(\Omega\text{loc})$, $\Gamma_2(\Omega\text{loc})$ or $C^\infty(\Omega)$, respectively, where Ω is an appropriate neighborhood of P.

Theorem 3.20 is proved in [11] as a consequence of the identity (3.19). The corollary follows from [9] if we note that K is a homogeneous distribution on H_n of degree $-2n$.

Observe that when $n=1$, then $v = -\bar{Z}_1(u)$ is a solution of Lewy's equation

(3.22) $$L(v) = (\frac{\partial}{\partial z} + i\bar{z}\frac{\partial}{\partial t})(v) = f$$

if $\square_b^{(0)}(u) = f$. Thus we have

3.23 Theorem.* Given f, then equation (3.22) has a solution in a neighborhood of a point $P \in \mathbb{R}^3$ if and only if the Cauchy-Szego integral $C_b(f)$ is real-analytic in a neighborhood of P.

3.24 Corollary. If the above condition for f is satisfied and if f belongs to one of the spaces $S_k^p(\Omega\text{loc})$, $\Gamma_\alpha(\Omega\text{loc})$, $L^\infty(\Omega\text{loc})$ or $C^\infty(\Omega)$, then we can find a v which belongs to $S_{k+1}^p(\Omega\text{loc})$, $\Gamma_{\alpha+1}(\Omega\text{loc})$, $\Gamma_1(\Omega\text{loc})$ or $C^\infty(\Omega)$, respectively, for appropriate neighborhoods Ω of P.

Actually, the necessity part of the condition of Theorem 3.23 gives the following statement: $L(v) = f$ has no solution in any neighborhood of the point P if $C(f)$ cannot be continued analytically across P; in particular, if f is the boundary value of a (suitably bounded) holomorphic function in $\mathfrak{D}$ which has no analytic extension across P.**

The next result "explains" the existence and derivation of the "relative fundamental solution" for $\square_b^{(0)}$. We define Ψ_α by

(3.25) $$\Psi_\alpha = \frac{\Gamma(\frac{n+\alpha}{2})\Gamma(\frac{n-\alpha}{2})}{2^{2-n}\pi^{n+1}}(|z|^2 - it)^{-\frac{n+\alpha}{2}}(|z|^2 + it)^{-\frac{n-\alpha}{2}}$$

$$- \frac{2^{n-1}\Gamma(n)}{(n-\alpha)\pi^{n+1}}(|z|^2 - it)^{-n}$$

*See the work of Sato, Kawai, and Kashiwara [30], for a related result.
**For this see Greiner, Kohn, and Stein [11].

and set $M_\alpha(f) = f * \Psi_\alpha$. Then we have

3.26 Proposition. (i) $\mathcal{L}_\alpha \circ M_\alpha = I - C_b$ for all α in some neighborhood of $\alpha = n$.

(ii) $\lim_{\alpha \to n} \Psi_\alpha = \Phi$, where the limit is taken pointwise, and in the sense of distributions.

Proof. (i) follows from a simple computation. As for (ii), we write

$$\Psi_\alpha = \frac{2^{n-2}\,\Gamma(n)}{\pi^{n+1}} (|z|^2 - it)^{-n}$$

$$\cdot \frac{1}{n-\alpha} \left(\frac{2\Gamma(\frac{n+\alpha}{2})}{\Gamma(n)} \Gamma(\tfrac{n-\alpha}{2}+1) \left(\frac{|z|^2 - it}{|z|^2 + it} \right)^{\frac{n-\alpha}{2}} - 2 \right)$$

$$\xrightarrow{\alpha \to n} -\frac{2^{n-2}\,\Gamma(n)}{\pi^{n+1}} (|z|^2 - it)^{-n} \left[\frac{d}{d\alpha} \left(\frac{|z|^2 - it}{|z|^2 + it} \right)^{\frac{n-\alpha}{2}} \right]_{\alpha=n}$$

$$= \Phi.$$

This completes the proof.

For future reference we state

3.27 Proposition. The symbol, $\sigma(C_b)$, of C_b is given by

$$\sigma(C_b) = 2^n e^{-|\omega|^2/\tau} \quad \text{if } \tau > 0,$$

$$= 0 \quad \text{if } \tau < 0,$$

where $\omega = (\omega_1, \ldots, \omega_n)$ is first used in (1.11).

This follows easily by taking the Fourier transform of the function $(\varepsilon^2 + |z|^2 - it)^{-n-1}$, and using (3.11).

Part II. Parametrix for the $\bar{\partial}$-Neumann problem

Guide to Part II

The purpose of this part is to reduce the $\bar{\partial}$-Neumann problem to the inversion of a pseudo-differential operator $\Box^+$ defined on the boundary; thereby one obtains a parametrix, an approximate "Neumann" operator, which gives a solution for our original problem, modulo controllable error terms.

Since the gist of the method is to reduce the question to the boundary, bM, and there to approximate by the Heisenberg group, certain preliminary problems must be dealt with, which we now describe.

Admissible coordinates

Under the assumption that bM is strongly pseudo-convex we can introduce, for each fixed $\xi \in bM$, a basic mapping $\eta \longrightarrow u(\xi, \eta)$ of a neighborhood in bM (centered at ξ) to a neighborhood of the origin in the Heisenberg group. We can view this mapping as giving an admissible coordinate system. It plays many roles. For example, in this coordinate system a basis of the vector fields in $T^{(1,0)}(bM)$ is well approximated (at ξ) by the corresponding standard basis in the Heisenberg group; see Proposition (4.3). Moreover if $f \longrightarrow \int_{H_n} k(y^{-1}x) f(y)\,dy$ is the exact form of an operator in the Heisenberg group derived from the complex structure, the operator

$$f \longrightarrow \int_{bM} k(u(\xi, \eta)) f(\xi)\,d\xi$$

is the corresponding (approximate) operator for bM. (See e.g., (9.6) and (10.6).) These ideas were already used in a crucial way in [9] in studying $\square_b$ on bM.

Levi metric

Another necessary preliminary is the construction of a Hermitian metric on M, which when restricted to $T^{1,0}(bM)$ agrees with a Levi form. All computations (of adjoints, etc.) are done with this metric. With this metric the operator $\square$ has a form compatible with the operators $\mathcal{L}_\alpha$, $\alpha = n-2$, on the Heisenberg group; see (6.23).

Exact form of $\square$

Since we are dealing with a non-elliptic problem, the first-order terms of $\square$ will also be critical. Let us describe the matter in more detail. One chooses first an orthonormal basis of the holomorphic vector fields $Z_1, Z_2, \ldots, Z_{n+1}$, so that $Z_1, \ldots, Z_n$ are tangential at bM, and which satisfy the commutation relations (6.20), where $Z_{n+1} = \frac{1}{\sqrt{2}}(\frac{\partial}{\partial\rho} + iT)$ and ρ denotes the geodesic distance from bM. According to our experience with the Heisenberg group, T is as significant as a quadratic expression in the Z_i and $\bar{Z}_j$, $1 \leq i, j \leq n$; but first-order terms in the Z_i (or $\bar{Z}_i$) $1 \leq i \leq n$, (i.e., $\varepsilon(Z, \bar{Z})$), are negligible. However Z_{n+1} plays a separate role. First-order terms in $\bar{Z}_{n+1}$ are negligible because of the second boundary condition (6.16). First-order terms in Z_{n+1} are not negligible but the exact form in (6.23) is ultimately balanced by the second boundary condition, since both are expressed in terms of the matrix S_{n+1}. The term $M_{n+1}(\varphi)$ in (6.23) is negligible because it involves essentially only the component $\nu(\varphi) = \varphi \lrcorner \bar{\omega}_{n+1}$, in the splitting $\varphi = \varphi_t + \bar{\omega}_{n+1} \nu(\varphi)$, and the first boundary condition (6.10) leads to an elliptic problem for this component.

Solution of a Dirichlet problem

The $\bar{\partial}$-Neumann problem is the following: To solve $\square u = f$, with the two boundary conditions $\nu(u) = 0$, and $\nu(\bar{\partial} u) = 0$ on bM, where $u = u_t + \bar{\omega}_{n+1} \nu(u)$. Solution of a standard Dirichlet problem reduces this to the determination of u_t on the boundary, and so the question then becomes the control of u_t on the boundary. The Dirichlet problem is solved, using known techniques, by the construction of a Green's operator G and a Poisson operator P. But here again we must keep track of the symbols which are one order lower than the main terms, and so only terms corresponding to order two lower are a priori negligible. This requires some lengthy analysis which is carried out in Chapter 7.

The operator $\square^+$

Once the Dirichlet problem is solved one turns to the determination of u_t on the boundary. It is then a question of inverting the operator $\square^+$, defined by $\square^+ = B_{\bar{\partial}} P$, where P is the Poisson integral, and $B_{\bar{\partial}}$ is the boundary operator coming from the second boundary condition. $\square^+$ is a pseudo-differential operator of order 1 (defined on bM), and in determining its symbol one must keep track of zero-order terms.

The operator $\square^+$ is then inverted by constructing another operator,

$\square^-$, similar to $\square^+$, so that $\square^- \square^+ = -\square_b$ approximately, where $\square_b$ is the Laplacian for (0,1) forms on the boundary. Now when $n > 1$, then $\square_b$ has a known parametrix K, and so the approximate inverse to $\square^+$ turns out to be $-K\square^-$.

The case n=1

When n=1, $\square_b$ is not invertible. This fact is intimately connected with the non-solvability of the Lewy equation. However using the results of Chapter 3, one can construct an operator $\overline{K}$, so that $\overline{K}\square_b = I - \overline{C}_b$ approximately, where $\overline{C}_b$ is essentially the projection operator on the boundary values of anti-holomorphic functions. Now away from its characteristic variety $\square^+$ is elliptic, and hence invertible. We can also write $I = E^+ + E^-$, where E^+ is an ordinary pseudo-differential operator of order 0 whose symbol is =1 near the characteristic variety of $\square^+$. The "projections" E^+ and $\overline{C}_b$ are approximately orthogonal since their symbols (respectively of types $S^0_{1,0}$, and $S^0_{1/2,1/2}$) have disjoint support. Thus an approximate inverse of $\square^+$ is given by $-K^+$, where $K^+ = E\,\overline{K}\square^- - E^-Q_-$, with Q_- an inverse of $\square^+$ away from its characteristic variety, i.e., $Q_-\square^+ = E^-$ approximately.

The approximate Neumann operator

As a result of the above analysis we get that the approximate Neumann operators (the inverse to our original problem) is given by (9.23), (or (10.29)).

The regularity properties of the various operators that make up the Neumann operator are the subject of Part III.

Chapter 4. Admissible coordinates on strongly pseudo-convex CR manifolds

Let $\mathfrak{M}$ be a CR manifold, i.e., a real oriented C^∞ manifold of dimension $2n+1$, $n=1,2,3,\ldots$, together with a subbundle $T^{1,0}$ of the complexified tangent bundle $\mathbb{C}T\mathfrak{M}$ satisfying

(a) $\dim_{\mathbb{C}} T^{1,0} = n$,

(b) $T^{1,0} \cap \overline{T}^{1,0} = \{0\}$,

(c) $T^{1,0}$ is integrable in the sense of Frobenius, i.e., if Z_1, Z_2 are sections of $T^{1,0}$, then so is their Lie bracket $[Z_1, Z_2]$.

Now we restrict our attention to a local coordinate patch U. Let $V_0, V_1, \ldots, V_{2n}$ denote a basis for the tangent bundle $T\mathfrak{M}$ on U, such that $Z_1, \ldots, Z_n$ yield a basis for $T_{1,0}$, where

(4.1) $$Z_j = \frac{1}{2}(V_j - iV_{j+n}), \quad j=1,\ldots,n.$$

If $\mathfrak{M}$ is strongly pseudo-convex we may assume, as we shall see later, that Z_j, $j=1,\ldots,n$ have been chosen to satisfy the following commutation relations.

(4.2) $$[Z_j, \overline{Z}_k] = -2i\,\delta_{jk} V_0 + \varepsilon(Z,\overline{Z}),$$

$j,k=1,\ldots,n$ where $\varepsilon(Z,\overline{Z})$ denotes "error terms," i.e., $\varepsilon(Z,\overline{Z}) \in T^{1,0} \oplus \overline{T}^{1,0}$

Via the exponential mapping Folland and Stein [9] constructed normal coordinates in U and showed, that in these coordinates Z_j, $j=1,\ldots,n$ are equal to their analogues on the Heisenberg group modulo "error terms."

See [9], §14. In the same paper they gave a more direct "geometric" construction for the normal coordinates in the case when $\mathfrak{M}$ is the strongly pseudo-convex boundary of a complex analytic manifold; see [9], §18.

The purpose of this chapter is to show that the "geometric" construction of normal coordinates can be carried out on arbitrary strongly pseudo-convex CR manifolds. Here is the main result.

4.3 Proposition. Let U be a coordinate neighborhood on $\mathfrak{M}$ with coordinates $w_0, w_1, \ldots, w_{2n}$. There exists a smooth mapping

$$(4.4) \qquad u: (\mathfrak{M} \cap U) \times (\mathfrak{M} \cap U) \longrightarrow \mathbb{R}_{2n+1},$$

such that, if $\xi, \eta \in \mathfrak{M} \cap U$ and

$$u(\xi, \eta) = (u_0(\xi, \eta), u_1(\xi, \eta), \ldots, u_{2n}(\xi, \eta)),$$

then we can write

$$(4.5) \qquad u_j(\xi, \eta) = \sum_{k=0}^{2n} L_j^{(k)}(\xi)(w_k(\eta) - w_k(\xi))$$

$$+ \sum_{k,\ell=0}^{2n} Q_j^{(k,\ell)}(\xi)(w_k(\eta) - w_k(\xi)(w_\ell(\eta) - w_\ell(\xi)),$$

$j=0, 1, \ldots, 2n$ and u has the following property: fix $\xi \in \mathfrak{M}$. Then $(u_0(\xi, \eta), u_1(\xi, \eta), \ldots, u_{2n}(\xi, \eta))$ gives a coordinate representation of $\eta \in \mathfrak{M} \cap U$. Then in this coordinate system (which is centered at ξ) we can write

$$(4.6) \qquad V_j = \frac{\partial}{\partial u_j} + 2u_{j+n} \frac{\partial}{\partial u_0} + \sum_{k=1}^{2n} O^1 \frac{\partial}{\partial u_{jk}} + O^2 \frac{\partial}{\partial u_0},$$

$$(4.7) \qquad V_{j+n} = \frac{\partial}{\partial u_{j+n}} - 2u_j \frac{\partial}{\partial u_0} + \sum_{k=1}^{2n} O^1 \frac{\partial}{\partial u_k} + O^2 \frac{\partial}{\partial u_0}, \quad j=1, \ldots, n, \text{ and}$$

$$(4.8)\qquad V_0 = \frac{\partial}{\partial u_0} + \sum_{j=0}^{2n} O^1 \frac{\partial}{\partial u_j},$$

where we used the notation

$$O^k = O(|u(\xi,\eta)|^k), \quad k=0,1,2,\ldots$$

with

$$(4.9)\qquad |u(\xi,\eta)| = |u_0(\xi,\eta)|^{1/2} + \sum_{j=1}^{2n} |u_j(\xi,\eta)|.$$

Proof. Let

$$(4.10)\qquad V_j = \sum_{k=0}^{2n} A_{jk}(\eta) \frac{\partial}{\partial w_k}, \quad j=0,1,\ldots,2n$$

Fix $\xi \in \mathcal{M}$ as a base point. We introduce new coordinates v_j, $j=0,1,\ldots,2n$ as follows

$$(4.11)\qquad v_j(\xi,\eta) = \sum_{k=0}^{2n} B_{jk}(\xi)(w_k(\eta) - w_k(\xi)),$$

$j=0,1,\ldots,2n$, where $B(\xi)^t = A(\xi)^{-1}$, $A = (A_{jk})$. Now we have

$$A_{jk}(\eta) = A_{jk}(\xi) + \sum_{\ell=0}^{2n} \frac{\partial A_{jk}}{\partial v_\ell}(\xi) v_\ell(\xi,\eta)$$

$$+ O\left(\sum_{j=0}^{2n} |v_j(\xi,\eta)|^2\right).$$

Consequently we have

$$V_j = \sum_{k=0}^{2n} \left\{A_{jk}(\xi) + \sum_{\ell=0}^{2n} \frac{\partial A_{jk}}{\partial v_\ell}(\xi) v_\ell(\xi,\eta)\right.$$

$$\left. + O\left(\sum_{i=0}^{2n} |v_i(\xi,\eta)|^2\right)\right\} \frac{\partial}{\partial w_k},$$

$j=0,1,\ldots,2n$. We substitute

$$\frac{\partial}{\partial w_k} = \sum_{i=0}^{2n} \frac{\partial v_i}{\partial w_k} \frac{\partial}{\partial v_i} = \sum_{i=0}^{2n} B_{ik}(\xi) \frac{\partial}{\partial v_i},$$

k=0, 1, . . . , 2n, and obtain

$$V_j = \frac{\partial}{\partial v_j} + \sum_{i,\ell=0}^{2n} \left(\sum_{k=0}^{2n} B_{ik}(\xi) \frac{\partial A_{jk}}{\partial v_\ell}(\xi) \right) v_\ell \frac{\partial}{\partial v_i}$$

$$+ O\left(\sum_{i=0}^{2n} |v_i(\xi, \eta)|^2 \right) \frac{\partial}{\partial v_i}$$

For the sake of simplicity we set

(4.12) $$a_{ijk}(\xi) = \sum_{\nu=0}^{2n} B_{k\nu}(\xi) \frac{\partial A_{i\nu}}{\partial v_j}(\xi),$$

i, j, k=0, 1, . . . , 2n. Then we can write

(4.13) $$V_0 = \frac{\partial}{\partial v_0} + \sum_{k=0}^{2n} O(|v|) \frac{\partial}{\partial v_k}, \quad \text{and}$$

(4.14) $$V_i = \frac{\partial}{\partial v_i} + \sum_{j,k=0}^{2n} a_{ijk}(\xi) v_j(\xi, \eta) \frac{\partial}{\partial v_k}$$

$$+ \sum_{k=0}^{2n} O(|v|^2) \frac{\partial}{\partial v_k},$$

i=1, . . . , 2n. We introduce new variables $u_0, u_1, \ldots, u_{2n}$ as follows

(4.15) $$u_0(\xi, \eta) = v_0(\xi, \eta) + \frac{1}{2} \sum_{j,k=1}^{2n} d_{jk}(\xi) v_j(\xi, \eta) v_k(\xi, \eta),$$

(4.16) $u_j = v_j$, j=1, . . . , 2n. We assume that $d_{jk}(\xi) = d_{kj}(\xi)$, n, k=1, . . . , 2n

Then

$$v_0(\xi, \eta) = u_0(\xi, \eta) - \frac{1}{2} \sum_{j,k=1}^{2n} d_{jk}(\xi) v_j(\xi, \eta) v_k(\xi, \eta)$$

and

$$\frac{\partial}{\partial v_0} = \sum_{j=0}^{2n} \frac{\partial u_j}{\partial v_0} \frac{\partial}{\partial u_j} = \frac{\partial}{\partial u_0},$$

$$\frac{\partial}{\partial v_j} = \sum_{k=0}^{2n} \frac{\partial u_k}{\partial v_j} \frac{\partial}{\partial u_k} = \frac{\partial u_0}{\partial v_j} \frac{\partial}{\partial u_0} + \frac{\partial}{\partial u_j}$$

$$= \frac{\partial}{\partial u_j} + \left(\sum_{k=1}^{2n} d_{jk}(\xi) v_k(\xi, \eta) \right) \frac{\partial}{\partial u_0},$$

$j=1, \ldots, 2n$. Substituting this into $V_0, V_1, \ldots, V_{2n}$ we obtain

$$V_0 = \frac{\partial}{\partial u_0} + \sum_{k=0}^{2n} O^1 \cdot \frac{\partial}{\partial u_k},$$

$$V_j = \frac{\partial}{\partial u_j} + \sum_{k=1}^{2n} \left((d_{jk}(\xi) + a_{jk0}(\xi)) u_k(\xi, \eta) \right) \frac{\partial}{\partial u_0}$$

$$+ \sum_{k=1}^{2n} O^1 \cdot \frac{\partial}{\partial u_k} + O^2 \cdot \frac{\partial}{\partial u_0}, \quad j=1, \ldots, 2n.$$

The conclusions of the proposition require that we choose the $d_{jk}(\xi)$ so that

$$\sum_{k=1}^{2n} (d_{jk}(\xi) + a_{jk0}(\xi)) u_k = 2u_{j+n}$$

$$\sum_{k=1}^{2n} (d_{(j+n)k}(\xi) + a_{(j+n)k0}(\xi)) u_k(\xi, \eta) = -2u_j(\xi, \eta),$$

$j=1, \ldots, n$, that is

$$\sum_{\substack{k=1 \\ k \neq j+n}}^{2n} (d_{jk}(\xi) + a_{jk0}(\xi)) u_k(\xi, \eta)$$

$$+ (d_{j(j+n)}(\xi) + a_{j(j+n)0}(\xi) - 2) u_{j+n}(\xi, \eta) = 0,$$

$$\sum_{\substack{k=1\\k\neq j}}^{2n} (d_{(j+n)k}(\xi) + a_{(j+n)k0}(\xi))u_k(\xi,\eta)$$

$$+ (d_{(j+n)j}(\xi) + a_{(j+n)j0}(\xi) + 2)u_j(\xi,\eta) = 0,$$

$j=1,\ldots,n$. Now, in

$$[V_j + iV_{j+n}, V_p + iV_{p+n}]$$

$$= [V_j, V_p] - [V_{j+n}, V_{p+n}] + i([V_{j+n}, V_p] + [V_j, V_{p+n}])$$

the coefficient of $\frac{\partial}{\partial v_0}$ vanishes, because the space of anti-holomorphic vector fields is closed under brackets. Therefore (4.13) and (4.14) easily yield

$$(a_{pj0}(\xi) - a_{jp0}(\xi)) - (a_{(p+n)(j+n)0}(\xi) - a_{(j+n)(p+n)0}(\xi))$$

$$+ i\{(a_{p(j+n)0}(\xi) - a_{(j+n)p0}(\xi)) + (a_{(p+n)j0}(\xi) - a_{j(p+n)0}(\xi))\} = 0$$

Similarly, in

$$[V_j + iV_{j+n}, V_p - iV_{p+n}]$$

$$= [V_j, V_p] + [V_{j+n}, V_{p+n}] + i\{[V_{j+n}, V_p] - [V_j, V_{p+n}]\}$$

the coefficient of $\frac{\partial}{\partial v_0}$ is $8i\delta_{jp} + O^1(v)$. In other words

$$(a_{pj0}(\xi) - a_{jp0}(\xi)) + (a_{(p+n)(j+n)0}(\xi) - a_{(j+n)(p+n)0}(\xi))$$

$$+ i\{(a_{p(j+n)0}(\xi) - a_{(j+n)p0}(\xi)) - (a_{(p+n)j0}(\xi) - a_{j(p+n)0})\}$$

$$= 8i\delta_{jp}.$$

Therefore

$$a_{jp0}(\xi) = a_{pj0}(\xi),$$

$$a_{(p+n)(j+n)0}(\xi) = a_{(j+n)(p+n)0}(\xi),$$

for $j=1,\ldots,n$, and

$$a_{p(j+n)0}(\xi) - a_{(j+n)p0}(\xi) = 4\,\delta_{jp},$$

$1 \leqq j, p \leqq n$. From the above we obtain

$$d_{jk}(\xi) = -a_{jk0}(\xi), \quad k \neq j+n$$

$$d_{(j+n)k}(\xi) = -a_{(j+n)k0}(\xi), \quad k \neq j,$$

$j,k=1,\ldots,n$. Since $a_{jp0}(\xi) = a_{pj0}(\xi)$ as long as $p \neq j+n$ and $j \neq p+n$, $j,p=1,\ldots,n$, we can write

$$d_{jp}(\xi) = -\frac{1}{2}(a_{jp0}(\xi) + a_{pj0}(\xi)), \tag{4.17}$$

$j,p=1,\ldots,2n$, with $j \neq p+n$, $p \neq j+n$. On the other hand

$$a_{j(j+n)0}(\xi) = a_{(j+n)j0}(\xi) + 4,$$

and we need

$$d_{j(j+n)}(\xi) = 2 - a_{j(j+n)0}(\xi),$$
$$d_{(j+n)j}(\xi) = -2 - a_{(j+n)j0}(\xi).$$

This implies

$$d_{j(j+n)}(\xi) - d_{(j+n)j}(\xi) = 4 - (a_{j(j+n)0}(\xi) - a_{(j+n)j0}(\xi)) = 4 - 4 = 0,$$

$$\Rightarrow d_{j(j+n)}(\xi) = d_{(j+n)j}(\xi) = -\frac{1}{2}(a_{j(j+n)0}(\xi) + a_{(j+n)j0}(\xi)).$$

Thus, finally we have

$$d_{jk}(\xi) = -\frac{1}{2}(a_{jk0}(\xi) + a_{kj0}(\xi)), \tag{4.18}$$

$j,k=1,\ldots,2n$.

Now we are ready to derive (4.5). For $j=0,1,\ldots,2n$ the first-order terms in $u_j(\xi,\eta)$ are given by $v_j(\xi,\eta)$. Since

$$v_j(\xi,\eta) = \sum_{k=0}^{2n} B_{jk}(\xi)(w_k(\eta) - w_k(\xi))$$

we have

$$L_j^{(k)}(\xi) = B_{jk}(\xi), \quad j,k=0,1,\ldots,2n. \tag{4.19}$$

From (4.13) we may take

$$Q_j^{(k,\ell)}(\xi) = 0, \quad j=1,\ldots,2n, \quad \text{and} \quad k,\ell=0,1,\ldots,2n. \tag{4.20}$$

To find $Q_0^{(k,\ell)}$ we set

$$u_0(\xi,\eta) = v_0(\xi,\eta) + \frac{1}{2}\sum_{i,j=0}^{2n} d_{ij}(\xi)\, v_i(\xi,\eta)\, v_j(\xi,\eta) \tag{4.21}$$

which is equal to (4.15) modulo $O(|v_0|^2 + |v_0||v'|)$ where $v' = (v_1,\ldots,v_{2n})$. This change does not affect the previous computations and it will greatly simplify our final formulas. Thus we consider

$$\frac{1}{2}\sum_{i,j=0}^{2n} d_{ij}(\xi)\, v_i(\xi,\eta)\, v_j(\xi,\eta)$$

$$= \sum_{i,j=0}^{2n} d_{ij}(\xi) \sum_{k,\ell=0}^{2n} B_{ik}(\xi) B_{j\ell}(\xi)(w_k(\eta) - w_k(\xi)) \cdot (w_\ell(\eta) - w_\ell(\xi))$$

Therefore

$$Q_0^{(k,\ell)}(\xi) = \frac{1}{2}\sum_{i,j=0}^{2n} d_{ij}(\xi)\, B_{ik}(\xi)\, B_{j\ell}(\xi) \tag{4.22}$$

Now

$$d_{ij}(\xi) = -\frac{1}{2}(a_{ij0}(\xi) + a_{ji0}(\xi)),$$

$$a_{ij0}(\xi) = \sum_{\nu=0}^{2n} \frac{\partial A_{i\nu}(\xi)}{\partial v_j} B_{0\nu}(\xi)$$

$$= \sum_{\nu,\gamma=0}^{2n} B_{0\nu}(\xi) \frac{\partial A_{i\nu}}{\partial w_\gamma}(\xi) A_{j\gamma}(\xi)$$

$$= -\sum_{\nu,\gamma=0}^{2n} \frac{\partial B_{0\nu}}{\partial w_\gamma}(\xi) A_{i\nu}(\xi) A_{j\gamma}(\xi)$$

Thus

$$(4.23) \qquad Q_0^{(k,\ell)}(\xi) = \frac{1}{4} \sum_{i,j=0}^{2n} \sum_{\nu,\gamma=0}^{2n} \left(\frac{\partial B_{0\nu}}{\partial w_\gamma}(\xi) + \frac{\partial B_{0\gamma}}{\partial w_\nu} \right)$$

$$\cdot A_{i\nu}(\xi) A_{j\gamma}(\xi) B_{ik}(\xi) B_{j\ell}(\xi)$$

$$= \frac{1}{4} \left(\frac{\partial B_{0k}}{\partial w_\ell}(\xi) + \frac{\partial B_{0\ell}}{\partial w_k}(\xi) \right).$$

Thus the $Q_j^{(k,\ell)}$ are determined and we have proved Proposition 4.3

(4.24) Corollary. Let $z_j = u_j + iu_{j+n}$, $j=1,\ldots,n$. Then

$$Z_j = \frac{\partial}{\partial z_j} + i\bar{z}_j \frac{\partial}{\partial u_0} + \sum_{k=1}^{n} \left(O^1 \frac{\partial}{\partial z_k} + O^1 \frac{\partial}{\partial \bar{z}_k} \right)$$

$$+ O^2 \frac{\partial}{\partial u_0},$$

where we set $Z_j = \frac{1}{2}(V_j - iV_{j+n})$, $j=1,\ldots,n$.

Next we discuss the behavior of u when interchanging ξ and η. The resulting symmetry, or more precisely, the approximate version given below, plays a useful role in various estimates. See [9], §15.

(4.25) Lemma. We have

$$u_0(\xi, \eta) + u_0(\eta, \xi) = O^3, \tag{4.26}$$

$$u_j(\xi, \eta) + u_j(\eta, \xi) = O^2, \quad j=1, \dots, 2n. \tag{4.27}$$

Proof. First of all, if $j=1, \dots, 2n$, we have

$$u_j(\xi, \eta) + u_j(\eta, \xi) = v_j(\xi, \eta) + v_j(\eta, \xi)$$

$$= \sum_{k=0}^{2n} (B_{jk}(\xi) - B_{jk}(\eta))(w_k(\eta) - w_k(\xi))$$

$$= O(|w(\eta) - w(\xi)|^2) = O^2,$$

which proves (4.27). As for (4.26), using either (4.15) or (4.21) we obtain

$$u_0(\xi, \eta) + u_0(\eta, \xi)$$

$$= v_0(\xi, \eta) + \frac{1}{2} \sum_{j,k=1}^{2n} d_{jk}(\xi) v_j(\xi, \eta) v_k(\xi, \eta)$$

$$+ v_0(\eta, \xi) + \frac{1}{2} \sum_{j,k=1}^{2n} d_{jk}(\eta) v_j(\eta, \xi) v_k(\eta, \xi) + O^3$$

$$= v_0(\xi, \eta) + v_0(\eta, \xi)$$

$$+ \sum_{j,k=1}^{2n} d_{jk}(\xi) v_j(\xi, \eta) v_k(\eta, \xi) + O(|v|^3) + O^3$$

Next

$$v_0(\xi, \eta) + v_0(\eta, \xi)$$

$$= \sum_{k=0}^{2n} (B_{0k}(\xi) - B_{0k}(\eta))(w_k(\eta) - w_k(\xi))$$

$$= - \sum_{k=0}^{2n} (B_{0k}(\eta) - B_{0k}(\xi)) \Big(\sum_{i=0}^{2n} A_{ik}(\xi) v_i(\xi,\eta) \Big)$$

$$= - \sum_{k=0}^{2n} \Big(\sum_{\ell=0}^{2n} \frac{\partial B_{0k}}{\partial v_\ell}(\xi) v_\ell \Big) \Big(\sum_{i=0}^{2n} A_{ik}(\xi) v_i(\xi,\eta) \Big)$$

$$+ O(|v(\xi,\eta)|^3)$$

$$= - \sum_{i,\ell=1}^{2n} \Big(\sum_{k=0}^{2n} \frac{\partial B_{0k}}{\partial v_\ell}(\xi) A_{ik}(\xi) \Big) v_i(\xi,\eta) v_\ell(\xi,\eta)$$

$$+ O(|v'||v_0| + |v_0|^2 + |v'|^3)$$

where $v' = (v_1, \ldots, v_{2n})$. Now

$$B_{0k}(\xi) A_{ik}(\xi) = \delta_{0i}$$

$$\Longrightarrow \frac{\partial B_{0k}}{\partial v_\ell}(\xi) A_{ik}(\xi) + B_{0k}(\xi) \frac{\partial A_{ik}}{\partial v_\ell}(\xi) = 0,$$

hence

$$v_0(\xi,\eta) + v_0(\eta,\xi)$$

$$= \sum_{i,\ell=1}^{2n} \Big(\sum_{k=0}^{2n} B_{0k}(\xi) \frac{\partial A_{ik}}{\partial v_\ell}(\xi) \Big) v_i(\xi,\eta) v_\ell(\xi,\eta) + O^3 .$$

Thus if we want (4.26) to hold we require

$$a_{i\ell 0}(\xi) + a_{\ell i 0}(\xi) + d_{i\ell}(\xi) + d_{\ell i}(\xi) = 0$$

$$\Longrightarrow d_{i\ell}(\xi) = -\frac{1}{2} (a_{i\ell 0}(\xi) + a_{\ell i 0}(\xi)),$$

$i, \ell = 1, \ldots, 2n$, which is just (4.18). This proves Lemma (4.25).

<u>4.28 Corollary</u>. <u>If</u> $\xi, \eta \in \mathfrak{M} \cap U$ <u>and</u> U <u>is sufficiently small, then</u>

$$|u_0(\xi,\eta)| + \sum_{j=1}^{2n} |u_j(\xi,\eta)|^2 \text{ is equivalent to } |u_0(\eta,\xi)| + \sum_{j=1}^{2n} |u_j(\eta,\xi)|^2.$$

4.29 Definition. Let $V_0, V_1, \ldots, V_{2n}$ represent our fixed basis of $T\mathfrak{M}$ satisfying (4.2) in U. We shall say that $u(\xi,\eta) = (u_0(\xi,\eta), u_1(\xi,\eta), \ldots$ $a_{2n}(\xi,\eta))$ is an admissible coordinate system in U centered at ξ, whenever $V_j, j=0,1,\ldots,2n$ expressed in the coordinates $u(\xi,\eta)$ has the form (4.6), (4.7) and (4.8).

4.30 Proposition. *All admissible coordinate systems are equivalent in the sense that, if* U *is sufficiently small and* $e(\xi,\eta)$ *is another admissible coordinate system, then*

(i) $e_0(\xi,\eta) = u_0(\xi,\eta) + O^3(u)$,

$e_j(\xi,\eta) = u_j(\xi,\eta) + O^2(u), \quad j=1,\ldots,2n,$

(ii) *There exist* $c_1, c_2 > 0$ *such that* $c_1\rho_u(\xi,\eta) \leqq \rho_e(\xi,\eta) \leqq c_2\rho_u(\xi,\eta)$, *where we set*

$$\rho_u(\xi,\eta) = |u_0(\xi,\eta)|^{1/2} + \sum_{j=1}^{2n} |u_j(\xi,\eta)|,$$

similarly for $\rho_e(\xi,\eta)$.

Proof. (i) $\Rightarrow$ (ii) so we have to prove (i). According to the hypotheses we have

$$e_k(\xi,\eta) = u_k(\xi,\eta) + \frac{1}{2}\sum_{p,q=0}^{2n} b_{kpq}(\xi)u_p(\xi,\eta)u_q(\xi,\eta)$$
$$+ O(|u|^3), \quad k=0,1,\ldots,2n.$$

Therefore

$$\frac{\partial}{\partial u_j} = \frac{\partial}{\partial e_j} + \sum_{k=0}^{2n} \left(\sum_{q=0}^{2n} b_{kjq}(\xi) u_q(\xi,\eta) \right) \frac{\partial}{\partial e_k}$$

$$+ O(|u|^2) \frac{\partial}{\partial e} ,$$

$j=0,1,\ldots,2n$, where $\partial/\partial e$ stands for any derivative with respect to the e coordinates. To continue

$$V_j = \frac{\partial}{\partial e_j} + \sum_{q=1}^{2n} b_{0jq}(\xi) u_q(\xi,\eta) \frac{\partial}{\partial e_0}$$

$$+ 2u_{j+n}(\xi,\eta) \frac{\partial}{\partial e_0}$$

$$+ \sum_{k=1}^{2n} O(|u|^2) \frac{\partial}{\partial e_k} + O^2 \cdot \frac{\partial}{\partial e_0} , \quad j=1,\ldots,n.$$

Since e is admissible

$$V_j = \frac{\partial}{\partial e_j} + 2e_{j+n} \frac{\partial}{\partial e_0} + \sum_{\ell=1}^{2n} O^1 \cdot \frac{\partial}{\partial e_\ell} + O^2 \cdot \frac{\partial}{\partial e_0} ,$$

and since

$$u_{j+n}(\xi,\eta) = e_{j+n}(\xi,\eta) + O(|e|^2),$$

$j=1,\ldots,n$, we have

$$b_{0jq}(\xi) = 0, \quad j=1,\ldots,n; \ q=1,\ldots,2n.$$

A similar computation shows that

$$b_{0jq}(\xi) = 0, \quad j=n+1,\ldots,2n; \quad q=1,\ldots,2n.$$

This proves (i) of Proposition 4.30.

An important set of coordinates are the normal coordinates, which are defined in an invariant fashion, and constitute an admissible coordinate system. To be more precise, fix $\xi \in \mathcal{M}$ and let E_ξ denote the <u>exponential</u>

<u>map</u> at ξ based on the frame $\{V_j\}$. That is, for $u = (u_0, u_1, \ldots, u_{2n+1})$ sufficiently close to the origin in $\mathbb{R}^{2n+1}$ define $E_\xi(u) \in \mathfrak{M}$ to be the endpoint $\eta(1)$ of the integral curve $\eta(s)$, $(0 \leqq s \leqq 1)$ of the vector field $\sum_{j=0}^{2n} u_j V_j$ with $\eta(0) = \xi$. Then, according to [9], §14, the normal coordinate mapping E_ξ^{-1}, in some sufficiently small neighborhood of ξ, is admissible. From our Proposition 4.30 and Theorem 14.10(d) of [9] we immediately have

<u>(4.31) Proposition.</u> <u>Let</u> ξ, η, ζ <u>vary over a sufficiently small neighborhood</u> U <u>in</u> $\mathfrak{M}$. <u>Let</u> u <u>be an admissible coordinate system. Then there exists constants</u> $c_1, c_2 > 0$ <u>such that</u>

$$|u(\xi,\eta) - u(\zeta,\eta)| \leqq c_1(\rho_u(\xi,\zeta) + \rho_u(\xi,\zeta)^{1/2}\rho_u(\xi,\eta)^{1/2}), \tag{4.32}$$

$$\rho_u(\zeta,\eta) \leqq c_2(\rho_u(\xi,\zeta) + \rho_u(\xi,\eta)). \tag{4.33}$$

Finally, we compute the volume element. Suppose we are given a metric on $\mathfrak{M}$ such that $Z_j = \frac{1}{2}(V_j - iV_{j+n})$, $j=1,\ldots,n$ is an orthonormal basis for $T_{1,0}(\mathfrak{M} \cap U)$, $U = U(y_0, y_1, \ldots, y_{2n})$ some coordinate neighborhood, i.e.,

$$\langle Z_j, Z_k\rangle = \delta_{jk}, \qquad j,k=1,\ldots,n, \tag{4.34}$$

$$\langle Z_j, \overline{Z}_k\rangle = 0, \qquad j,k=1,\ldots,n, \tag{4.35}$$

$$\langle V_0, V_j\rangle = \delta_{0j}, \qquad j=0,1,\ldots,2n. \tag{4.36}$$

Set

$$V_j = \sum_{p=0}^{2n} A_{jp}(y)\frac{\partial}{\partial y_p}.$$

Then (4.34) and (4.35) yield

$$\frac{1}{4}(\langle V_j, V_k\rangle + \langle V_{j+n}, V_{k+n}\rangle)$$

$$+ i\frac{1}{4}(\langle V_j, V_{k+n}\rangle - \langle V_{j+n}, V_k\rangle) = \delta_{jk},$$

and

$$\frac{1}{4}(\langle V_j, V_k\rangle - \langle V_{j+n}, V_{k+n}\rangle)$$

$$- i\frac{1}{4}(\langle V_j, V_{k+n}\rangle - \langle V_{j+n}, V_k\rangle) = 0,$$

$j, k=1, 2, \ldots, n$. Consequently we have

$$\frac{1}{2}\langle V_j, V_k\rangle = \delta_{jk}, \quad j, k=1, 2, \ldots, 2n.$$

Hence $v_0, v_1, \ldots, v_{2n}$ is an orthonormal basis for $T\mathfrak{M}$, where we set $v_0 = V_0$ and $v_j = \frac{1}{\sqrt{2}} V_j$, $j=1, \ldots, 2n$. Set

$$v_j = \sum_p a_{jp}(y)\frac{\partial}{\partial y_p}, \quad j=0, 1, \ldots, 2n.$$

Then

$$\delta_{jk} = \langle v_j, v_k\rangle$$

$$= \langle \sum_{p=0}^{2n} a_{jp}(y)\frac{\partial}{\partial y_p}, \sum_{q=0}^{2n} a_{kq}(y)\frac{\partial}{\partial y_q}\rangle$$

$$= \sum_{p,q=0}^{2n} a_{jp}(y)\, g_{pq}(y)\, a_{kq}(y),$$

where we set $\langle\frac{\partial}{\partial y_p}, \frac{\partial}{\partial y_q}\rangle(y) = g_{pq}(y)$. In other words $a(y)\,g(y)\,a(y)^t = I$, or $\det g(y) = (\det a(y))^{-2}$. Thus we have derived

__4.37 Proposition.__ _Suppose we are given a metric on $\mathfrak{M}$ such that_ V_0

and $Z_j = \frac{1}{2}(V_j - iV_{j+n})$, $j=1,\ldots,n$ *satisfy* (4.34), (4.35) *and* (4.36). *Set*

(4.38)
$$V_j = \sum_{p=0}^{2n} A_{jp}(y) \frac{\partial}{\partial y_p}$$

in some coordinate neighborhood $U(y_0, y_1, \ldots, y_{2n})$. *Then*

(4.39)
$$d\mathfrak{m} = 2^{-n} |\det A(y)|^{-1} dy.$$

Chapter 5. Levi metrics

Let M be a sub-domain with smooth boundary bM of a complex manifold M′. Then to each P point of bM one can assign a Levi-form, a Hermitian form on $T^{1,0}(bM)|_P$. (This Levi-form is not unique, but is determined up to a positive multiple.) The assumption that M is strongly pseudo-convex means that this form is strictly positive definite at each point $P \in bM$. The purpose of this chapter is to give an explicit construction of a Hermitian metric on M, which restricted to $T^{1,0}(bM)$ coincides with the Levi form (i.e., (5.7) is satisfied). We call such a metric a Levi metric. We always work with such a metric because the boundary theory in [9] requires it.*

Let r' be a C^∞ real-valued function on an n+1 dimensional complex analytic manifold M′ and let $M = \{r' > 0\} \subset M'$ be a relatively compact domain with boundary bM. If $dr' \neq 0$ on bM then bM is a 2n+1 dimensional real C^∞ manifold. We shall assume that M is strongly pseudo-convex, that is $\{r'_{z_i \bar{z}_j}\}_{i,j=1,\dots,n+1}$ is positive definite at every point P of bM on vectors $(a_1, \dots, a_{n+1})$ which satisfy the side condition

$$(5.1) \qquad \sum_{j=1}^{n+1} a_j r'_{z_j}(P) = 0.$$

This condition is invariant under holomorphic coordinate transformations. Here $(z_1, \dots, z_{n+1})$ denotes an arbitrary analytic coordinate system in some neighborhood of $P \in bM$. According to Proposition 4.4 of [21] there

---

*See however the remarks in the concluding section of the Introduction.

exists a positive number A, such that

$$r'' = e^{Ar'} - 1$$

induces a positive definite Hermitian metric, namely

(5.2) $$ds^2 = \sum_{i,j=1}^{n+1} r''_{z_i\bar{z}_j} dz_i d\bar{z}_j$$

in some neighborhood V of bM. (This is only a preliminary metric, not the final one constructed.) Let $\langle\ ,\ \rangle_{r''}$ denote the inner product induced by this metric on the complexified tangent bundle $T^{(1,0)}(V) \oplus T^{(0,1)}(V)$ of V. Choose a neighborhood U of bM, $\overline{U} \subset V$ and a positive C^∞ function φ on M' such that $\varphi \equiv 1$ on $\overline{U}$ and φ vanishes outside of V. Let $\{h_{i\bar{j}}\}_{i,j=1,\ldots,n+1}$ denote an arbitrary positive definite Hermitian metric on M' and set

(5.3) $$g_{i\bar{j}} = \varphi \frac{r''_{z_i\bar{z}_j}}{2\langle \partial r'', \partial r''\rangle_{r''}} + (1-\varphi)h_{i\bar{j}},$$

$i,j=1,\ldots,n+1$. We shall show that the metric $\{g_{i\bar{j}}\}_{i,j=1,\ldots,n+1}$ reduces to the Levi form on bM.

Let $T^{(1,0)}(bM)$ denote the set of holomorphic vector fields tangent to bM, i.e., in analytic local coordinates $(z_1,\ldots,z_{n+1})$, $z_j = x_{2j-1} + ix_{2j}$, $j=1,\ldots,n+1$,

$$L = \sum_{j=1}^{n+1} a_j(x)\frac{\partial}{\partial z_j} \in T^{(1,0)}(bM)$$

$$\Longleftrightarrow \sum_{j=1}^{n+1} a_j(x)\frac{\partial r''}{\partial z_j} = 0 \text{ on } bM.$$

The Levi form $\langle \, , \, \rangle_L$ is an Hermitian inner product on $T^{(1,0)}(bM)$ defined as follows. Set

(5.4) $$r = \frac{r''}{2\langle \partial r'', \partial r'' \rangle_{r''}} .$$

Clearly

$$\partial r = \frac{\partial r''}{2\langle \partial r'', \partial r'' \rangle_{r''}}$$

on bM. Let $L_1, L_2 \in T^{(1,0)}(bM)$. We define

(5.5) $$\langle L_1, L_2 \rangle_L = \frac{i}{2}(i(\bar{\partial} r - \partial r), [L, \bar{L}])$$

where the right-hand side denotes the usual contraction operation between tangent and cotangent vectors and $[L_1, \bar{L}_2] = L_1\bar{L}_2 - \bar{L}_2 L_1$. We denote by $\langle \cdot , \cdot \rangle$ the inner product given by our metric (5.3).

5.6 Proposition. *Let* $L_1, L_2 \in T^{(1,0)}(bM)$. *Then*

(5.7) $$\langle L_1, L_2 \rangle (P) = \langle L_1, L_2 \rangle_L (P)$$

for all $P \in bM$. *We also have*

$$\langle \partial r, \partial r \rangle (P) = \frac{1}{2} ,$$

consequently

(5.8) $$\langle dr, dr \rangle (P) = \langle i(\bar{\partial} r - \partial r), i(\bar{\partial} r - \partial r) \rangle (P) = 1$$

for all $P \in bM$.

Proof. We choose an appropriate coordinate system in which to carry out our calculations. We may as well assume that the hyperplane $\operatorname{Im} z_{n+1} = 0$ is tangent to bM at P and P is the origin of the coordinate system. Thus

$$r'' = -y_{n+1} + A(z) + O(|z|^3),$$

where from Taylor's formula we have

$$A(z) = \sum_{j,k=1}^{n+1} z_j \bar{z}_k \frac{\partial^2 r''(0)}{\partial z_j \partial \bar{z}_k} + \mathrm{Re} \sum_{j,k=1}^{n+1} z_j z_k \frac{\partial^2 r''(0)}{\partial z_j \partial z_k} .$$

Next we make the following analytic change of variables

$$z'_k = z_k, \quad k=1, \ldots, n,$$

$$z'_{n+1} = z_{n+1} - i \sum_{j,k=1}^{n+1} z_j z_k \frac{\partial^2 r''(0)}{\partial z_j \partial z_k} .$$

In these coordinates r'' assumes the simpler form

$$r''(z') = -\mathrm{Im}\, z'_{n+1} + \sum_{j,k=1}^{n+1} z'_j \overline{z'_k} \frac{\partial^2 r''(0)}{\partial z'_j \partial \bar{z}'_k} + O(|z|^3)$$

Thus we may as well assume that r'' is of the form

$$\text{(5.9)} \qquad r'' = -\mathrm{Im}\, z_{n+1} + \sum_{j,k=1}^{n+1} A_{j\bar{k}} z_j \bar{z}_k + O(|z|^3)$$

where $(A_{j\bar{k}})$ is a positive definite Hermitian symmetric matrix. To prove (5.7) it suffices to compute the two sides at $P = (0, \ldots, 0)$ in the case that L_1 and L_2 belong to a basis of $T^{(1,0)}(bM)$. Locally such a basis is given by

$$p_j = r''_{z_{n+1}} \frac{\partial}{\partial z_j} - r''_{z_j} \frac{\partial}{\partial z_{n+1}} , \quad j=1, \ldots, n.$$

That this is a local basis follows by noting that p_j, $j=1, \ldots, n$ and $\frac{\partial}{\partial z_{n+1}}$ is a basis of $T^{(1,0)}(M')$ at P. Now at P, that is at $z = (0, \ldots, 0)$,

$p_j = \frac{1}{2} i \frac{\partial}{\partial z_j}$, $j=1,\ldots,n$, therefore

$$\langle p_j, p_k \rangle (P) = \frac{\frac{1}{4} A_{j\bar{k}}}{2\langle \partial r'', \partial r'' \rangle_{r''}(P)}$$

On the other hand a simple computation yields

$$[p_j, \bar{p}_k]_{(z=0)}$$

$$= \frac{1}{2} i A_{j(\overline{n+1})} \frac{\partial}{\partial \bar{z}_k} - \frac{1}{2} i A_{j\bar{k}} \frac{\partial}{\partial \bar{z}_{n+1}}$$

$$+ \frac{1}{2} i A_{k(\overline{n+1})} \frac{\partial}{\partial z_j} - \frac{1}{2} i A_{j\bar{k}} \frac{\partial}{\partial z_{n+1}} .$$

Here we evaluated the coefficients at the origin. Finally, at $z = 0$

$\partial r'' = \frac{1}{2} i dz_{n+1}$ and $\bar{\partial} r - \partial r = (-\frac{1}{2} i d\bar{z}_{n+1} - \frac{1}{2} i dz_{n+1}) \frac{1}{2\langle \partial r'', \partial r'' \rangle_{r''}(0)}$.

Therefore

$$\langle p_j, p_k \rangle_L (0) = \frac{1}{8} \frac{A_{j\bar{k}}}{\langle \partial r'', \partial r'' \rangle_{r''}(0)}$$

$$= \langle p_j, p_k \rangle (0)$$

Next we prove (5.8). To this end we note that if f is a form of type $(1,0)$ and $f = \sum_{j=1}^{n+1} f_j dz_j$ in local analytic coordinates, then

$$\langle f, f \rangle = \sum_{j,k=1}^{n+1} g^{j\bar{k}} f_j \bar{f}_k \tag{5.10}$$

where $(g^{j\bar{k}})$ is the inverse of $(g_{j\bar{k}})$, that is

$$g^{j\bar{k}} = 2\langle \partial r'', \partial r'' \rangle_{r''} r''^{j\bar{k}} \tag{5.11}$$

where, similarly, $(r''^{j\bar{k}})$ is the inverse of $(r''_{j\bar{k}})$. Therefore

$$\langle \partial r, \partial r \rangle (0) = \frac{1}{4\langle \partial r'', \partial r'' \rangle_{r''}(0)^2} \langle \partial r'', \partial r'' \rangle (0)$$

$$= \frac{1}{2},$$

and

$$\langle dr, dr \rangle (0) = \langle \partial r, \partial r \rangle (0) + \langle \partial r, \partial r \rangle (0)$$

$$= 1$$

$$= \langle i(\bar{\partial} r - \partial r), i(\bar{\partial} r - \partial r) \rangle (0),$$

and by the invariance under holomorphic coordinate changes we have proved Proposition (5.6).

Examples

We give two examples of Levi metrics.

(i) Let $\overline{M}$ be the unit ball in $\mathbb{C}^{n+1}$. Then the usual Euclidean metric is a Levi metric.

(ii) Let M be the domain $\operatorname{Im} z_{n+1} > |z_1|^2 + |z_2|^2 \dots + |z_n|^2$.

Then there is a Levi metric on $\overline{M}$ which has the following properties, among others: (a) The metric is invariant under the translations of M coming from the Heisenberg group H_n. (b) The distance of a point from the boundary is given by $\rho = \operatorname{Im} z_{n+1} - \sum_{j=1}^{n} |z_j|^2$. Further details are described in the example at the end of Chapter 17.

Chapter 6. $\square$ on (0,1)-forms

From now on we assume a fixed Hermitian metric $(g_{i\bar{j}})$ on M which satisfies (5.7) on bM, (a "Levi" metric). The following analysis is done in a fixed boundary coordinate neighborhood U. Let ρ denote geodesic distance from bM, at least in some sufficiently small neighborhood of bM, $\rho > 0$ in M and $\rho < 0$ outside of $\overline{M}$. Let

(6.1) $$\omega_1, \ldots, \omega_n, \omega_{n+1} = \sqrt{2}\,\partial\rho$$

denote an orthonormal basis for $T^{(1,0)}(M)^*$ in U and let $Z_1, \ldots, Z_{n+1}$ denote the dual basis for $T^{(1,0)}(U)$. We have

(6.2) $$df = \sum_{j=1}^{n+1} (Z_j f)\omega^j + \sum_{j=1}^{n+1} (\overline{Z}_j f)\overline{\omega}^j,$$

or, equivalently

(6.3) $$Z_j f = \langle df, \omega^j \rangle, \quad j=1, \ldots, n+1$$

for C^∞ functions f. (6.3) implies that in local coordinates $\overline{Z}_j$ is the complex conjugate of Z_j. Now, if $j=1, \ldots, n$, then

$$Z_j\rho = \langle d\rho, \omega^j \rangle = \langle \partial\rho, \omega^j \rangle = \frac{1}{\sqrt{2}} \langle \omega_{n+j}, \omega_j \rangle = 0,$$

and therefore $Z_j, \ldots, Z_n$ are tangential holomorphic vector fields. Furthermore so is $\frac{1}{\sqrt{2}} T = \operatorname{Im} Z_{n+1}$, because

(6.4) $$\frac{1}{2i}(Z_{n+1} - \overline{Z}_{n+1})\rho$$

$$= \frac{1}{2i} (\langle d\rho, \omega^{n+1} \rangle - \langle d\rho, \overline{\omega}^{n+1} \rangle)$$

$$= \frac{1}{\sqrt{2}\, i} (\langle \partial\rho, \partial\rho \rangle - \langle \overline{\partial}\rho, \overline{\partial}\rho \rangle)$$

$$= 0.$$

Finally

$$(6.5)\qquad (\mathrm{Re}\ Z_{n+1})f = \frac{1}{2}(Z_{n+1} + \overline{Z}_{n+1})f$$

$$= \frac{1}{2}(\langle df, \omega^{n+1}\rangle + \langle df, \overline{\omega}^{n+1}\rangle)$$

$$= \frac{1}{\sqrt{2}}(\langle df, \partial\rho\rangle + \langle df, \overline{\partial}\rho\rangle)$$

$$= \frac{1}{\sqrt{2}}\langle df, d\rho\rangle = \frac{1}{\sqrt{2}}\frac{\partial f}{\partial\rho},$$

since $\langle d\rho, d\rho\rangle = 1$.

We shall assume a boundary coordinate system $(\rho, x_1, \ldots, x_{2n+1})$ in U in which the metric has the form

$$ds^2 = \sum_{i,j=1}^{2n+1} g_{ij}\,dx^i dx^j + d\rho^2 .\text{*}$$

The volume element is

$$(6.6)\qquad dV = d\rho dS = \det(g_{ij})\,d\rho dx$$

where

$$(6.7)\qquad dS = \det(g_{ij})\,dx.$$

Next we compute the adjoints. Let u be a C_0^∞ function in U and

$$\varphi = \sum_{j=1}^{n+1} \varphi_j \overline{\omega}^j .$$

Then

$$\int_{\rho>0} \langle\overline{\partial}u, \varphi\rangle dV = \sum_{j=1}^{n+1}\int_{\rho>0} (\overline{Z}_j u)\overline{\varphi}_j\, dV$$

$$= \sum_{j=1}^{n+1}\int_{\rho>0} u(\overline{\overline{Z}_j^{*}\varphi_j})\, dV - \frac{1}{\sqrt{2}}\int_{\rho=0} u\overline{\varphi}_{n+1}\, dS,$$

---

*We point out that (g_{ij}) is not $(g_{i\overline{j}})$ of (5.3).

where we used (6.5) and set

$$Z_j^* = -\overline{Z}_j + \overline{h}_j, \quad j=1,\dots,n+1. \tag{6.8}$$

Thus we have

$$\vartheta\varphi = \sum_{j=1}^{n+1} (-Z_j + h_j)\varphi_j \tag{6.9}$$

and the first Neumann boundary condition is given by

$$(\nu\varphi)(bM) = \left(\overline{\omega}_{n+1} \rfloor \sum_{j=1}^{n+1} \varphi_j \overline{\omega}_j\right)(bM) = \varphi_{n+1}(bM) \tag{6.10}$$

where the interior product is defined as follows. Let $\overline{\omega}^J = \overline{\omega}_{j_1} \wedge \dots \wedge \overline{\omega}_{j_q}$ $J = (j_1, \dots, j_q)$. Then

$$\overline{\omega}^m \rfloor \overline{\omega}^J = 0 \quad \text{if } m \notin \{J\},$$

otherwise

$$\overline{\omega}^m \rfloor \overline{\omega}^J = (-1)^{i-1} \overline{\omega}_{j_1} \wedge \dots \wedge \overline{\omega}_{j_{i-1}} \wedge \overline{\omega}_{j_{i+1}} \wedge \dots \wedge \overline{\omega}_{j_q}$$

if $m = j_i$.

Next

$$\overline{\partial}\left(\sum_{j=1}^{n+1} \varphi_j \overline{\omega}^j\right) = \sum_{j<k} (\overline{Z}_j \varphi_k - \overline{Z}_k \varphi_j)\overline{\omega}^j \wedge \overline{\omega}^k + \sum_{j<k;\ell} \overline{s}_{jk}^{\ell} \varphi_\ell \, \overline{\omega}^j \wedge \overline{\omega}^k \tag{6.11}$$

where we set

$$\overline{\partial}\,\overline{\omega}^{\ell} = \sum_{j<k} \overline{s}_{jk}^{\ell} \, \overline{\omega}^j \wedge \overline{\omega}^k. \tag{6.12}$$

We note that

$$\bar{s}^{n+1}_{jk} = 0, \quad \text{because} \quad \bar{\partial}(\sqrt{2}\,\bar{\partial}\rho) = 0.$$

To compute the adjoint we set

$$\psi = \sum_{j<k} \psi_{jk}\,\bar{\omega}^j \wedge \bar{\omega}^k .$$

Then

$$\sum_{j<k} \int_{\rho>0} (\bar{Z}_j \varphi_k - \bar{Z}_k \varphi_j)\bar{\psi}_{jk}\, dV$$

$$= \sum_{j<k} \int_{\rho>0} \varphi_k\, \overline{(\bar{Z}_j^* \psi_{jk})}\; dV$$

$$- \sum_{j>k} \int_{\rho>0} \varphi_k\, \overline{(\bar{Z}_j^* \psi_{kj})}\, dV$$

$$+ \sum_{j=1}^{n} \frac{1}{\sqrt{2}} \int_{\rho=0} \varphi_j \bar{\psi}_{j,n+1}\, dS$$

Choosing $\varphi_j = \psi_{j,n+1}$, $j=1,\ldots,n$ we obtain the first Neumann boundary condition on $(0,2)$ forms, namely,

$$\psi_{j,n+1}(bM) = 0, \quad j=1,\ldots,n \tag{6.13}$$

To simplify the notation we set

$$\psi_{jk} = -\psi_{kj} \qquad \text{if } j>k$$

and $\psi_{jj} = 0$.

Similarly we set

$$s^{\ell}_{jk} = -s^{\ell}_{kj} \qquad \text{if } j \geqq k.$$

Then

$$(6.14)\qquad \vartheta\psi = \sum_{k=1}^{n+1}\Big(\sum_{j=1}^{n+1} \overline{Z}_j^{*}\,\psi_{jk}\Big)\overline{\omega}^k + \sum_{\ell=1}^{n}\Big(\sum_{j<k} s_{jk}^{\ell}\,\psi_{jk}\Big)\overline{\omega}^{\ell}.$$

Finally we shall translate the first Neumann boundary condition ν on (0,2) forms into the second Neumann boundary condition $\nu\overline{\partial}$ on (0,1) forms. This requires that (6.11) satisfies (6.13), that is

$$(6.15)\qquad (\nu\overline{\partial}\varphi)(bM) = \Big(\overline{Z}_j\varphi_{n+1} - \overline{Z}_{n+1}\varphi_j + \sum_{\ell=1}^{n} \overline{s}_{j(n+1)}^{\ell}\varphi_{\ell}\Big)(bM) = 0,$$

$j=1,2,\ldots,n$. If we assume that $\varphi = \sum_{j=1}^{n+1}\varphi_j\overline{\omega}^j$ already satisfies the first Neumann boundary condition $(\nu\varphi)(bM) = \varphi_{n+1}(bM) = 0$, then

$$(6.16)\qquad (\nu\overline{\partial}\varphi)(bM) = \Big(\overline{Z}_{n+1}\varphi_j - \sum_{\ell=1}^{n} \overline{s}_{j(n+1)}^{\ell}\varphi_{\ell}\Big)(bM) = 0,$$

$j=1,\ldots,n$, since Z_j, $j=1,\ldots,n$ are vector fields tangential to bM.

Next we compute the complex Laplacian $\square^{(1)} = \vartheta\overline{\partial} + \overline{\partial}\vartheta$ on (0,1)-forms. This is a lengthy but not difficult calculation. We shall only give the main steps. For simplicity of notation we shall from now on drop the superscript from $\square^{(1)}$ and write it as $\square$.

Let φ be a C^{∞} (0,1)-form. Then

$$\vartheta\overline{\partial}\Big(\sum_{j=1}^{n+1}\varphi_j\overline{\omega}^j\Big) = -\sum_{j=1}^{n+1}\Big(\sum_{i=1}^{n+1} Z_i[\overline{Z}_i\varphi_j - \overline{Z}_j\varphi_i]\Big)\overline{\omega}^j$$

$$+\left(S^{t}_{n+1}(\overline{Z}_{n+1}\varphi)\right)\overline{\omega}$$

$$+\sum_{j=1}^{n} h_{n+1}\left(\overline{Z}_{n+1}\varphi_j\right)\overline{\omega}^j$$

$$-\left(\sum_{j=1}^{n} h_j(\overline{Z}_{n+1}\varphi_j)\right)\overline{\omega}^{n+1}$$

$$-(\overline{S}_{n+1}(Z_{n+1}\varphi))\overline{\omega}$$

$$+\varepsilon(Z,\overline{Z})\varphi+\varepsilon(\varphi),$$

where

$$(S_{n+1})_{k,j}=s^{k}_{(n+1)j}, \tag{6.17}$$

where s^k_{ij} is defined by (6.12). $\varepsilon(Z,\overline{Z})$ stands for linear combinations of the vector fields $Z_i, \overline{Z}_j$, $i,j=1,\ldots,n$, (which are tangential at the boundary) with C^∞ coefficients and $\varepsilon(\varphi)$ stands for terms of order zero, i.e., multiplication by C^∞ functions. Similarly

$$\overline{\partial}\vartheta\left(\sum_{j=1}^{n+1}\varphi_j\overline{\omega}^j\right)$$

$$=-\sum_{i=1}^{n+1}\left(\sum_{j=1}^{n+1}\overline{Z}_i Z_j\varphi_j\right)\overline{\omega}^i$$

$$+\sum_{j=1}^{n+1} h_j(\overline{Z}_{n+1}\varphi_j)\overline{\omega}^{n+1}+\varepsilon(\overline{Z})(\varphi)+\varepsilon(\varphi).$$

h_j, $j=1,\ldots,n+1$ are defined in (6.8). Together these formulas yield $\Box$ as follows.

$$\square\Big(\sum_{j=1}^{n+1}\varphi_j\overline{\omega}^j\Big) = (\overline{\partial}\vartheta+\vartheta\overline{\partial})\Big(\sum_{j=1}^{n+1}\varphi_j\overline{\omega}^j\Big)$$

$$= \sum_{j=1}^{n+1}\Big\{\Big[-\frac{1}{2}\sum_{i=1}^{n}(Z_i\overline{Z}_i+\overline{Z}_iZ_i) - Z_{n+1}\overline{Z}_{n+1}$$

$$+ \Big([Z_j,\overline{Z}_j] - \frac{1}{2}\sum_{i=1}^{n}[Z_i,\overline{Z}_i]\Big)(\varphi_j)\Big\}\overline{\omega}^j$$

$$+ \sum_{j=1}^{n+1}\Big\{\sum_{\substack{i=1\\ i\neq j}}^{n+1}[Z_i,\overline{Z}_j](\varphi_i)\Big\}\overline{\omega}^j$$

$$+ \Big((h_{n+1}I_{n+1} + S^t_{n+1})\circ(\overline{Z}_{n+1}\varphi)\Big)\overline{\omega} - \Big(\overline{S}_{n+1}(Z_{n+1}\varphi)\Big)\overline{\omega}$$

$$+ \varepsilon(Z,\overline{Z})(\varphi) + \varepsilon(\varphi).$$

We are going to put $\square$ into diagonal form, modulo "remainder" terms, but first we make a few remarks We write

$$Z_{n+1} = \frac{1}{\sqrt{2}}\frac{\partial}{\partial\rho} + i\frac{1}{\sqrt{2}}T \tag{6.18}$$

where T is a real vector field. Then

$$\langle\frac{\sqrt{2}}{2i}(Z_{n+1} - \overline{Z}_{n+1}), \frac{\sqrt{2}}{2i}(Z_{n+1} - \overline{Z}_{n+1})\rangle$$

$$= \frac{1}{2}\langle Z_{n+1} - \overline{Z}_{n+1}, Z_{n+1} - \overline{Z}_{n+1}\rangle = 1,$$

therefore T has unit length. Next we note that T and $i(\partial\rho - \overline{\partial}\rho)$ are dual to each other, because

$$(i(\partial\rho - \overline{\partial}\rho), \frac{\sqrt{2}}{2i}(Z_{n+1} - \overline{Z}_{n+1}))$$

$$= \frac{1}{2}(\omega_{n+1} - \overline{\omega}_{n+1}, Z_{n+1} - \overline{Z}_{n+1})$$

$$= 1.$$

If r is defined by (5.4) then

(6.19) $$\bar{\partial} r - \partial r = \partial \rho - \bar{\partial} \rho$$

on bM, because both sides annihilate $T^{(0,1)}(bM) \oplus T^{(1,0)}(bM)$, both sides have unit length on bM --see (5.8)--and they point in opposite direction.

<u>(6.20) Lemma.</u> <u>For</u> $j, k=1, \ldots, n$ <u>we have</u>

$$[Z_j, \overline{Z}_k] = (-2i\,\delta_{jk} + O(\rho))\,T + \epsilon(Z, \overline{Z}).$$

<u>Proof.</u> Clearly

$$[Z_j, \overline{Z}_k] = c_{jk} T + \epsilon(Z, \overline{Z}),$$

and we have

(6.21) $$\begin{aligned} c_{jk} &= (i(\partial\rho - \bar{\partial}\rho), [Z_j, \overline{Z}_k]) \\ &= (i(\partial\rho - \bar{\partial}\rho), [Z_j, \overline{Z}_k])\,(bM) + O(\rho) \\ &= (i(\bar{\partial} r - \partial r), [Z_j, \overline{Z}_k])\,(bM) + O(\rho) \\ &= -2i\,\langle Z_j, Z_k\rangle\,(bM) + O(\rho) \\ &= -2i\,\delta_{jk} + O(\rho), \end{aligned}$$

where we used Proposition 5.6.

To put our formulas in final form let us recall our notation for "remainder terms."

S_{n+1} is a matrix of C^∞ functions, $(S_{n+1})_{k,j} = s^{(k)}_{(n+1)j}$ --see (6.12)-- $\epsilon(Z, \overline{Z})$ denotes a linear combination of $Z_i, \overline{Z}_j$, $i, j=1, \ldots, n$, with C^∞ coefficients, $O(\rho)$ stands for C^∞ functions which vanish on bM. We also let M_{n+1} stand for an $(n+1) \times (n+1)$ matrix of tangential differential operators, i.e., vector fields that do not include $\frac{\partial}{\partial\rho}$, such that M_{n+1} is zero except, possibly, in its $n+1^{\text{st}}$ column or $n+1^{\text{st}}$ row.

The term M_{n+1} comes from the terms $[Z_i, \overline{Z}_j]\varphi_i$, $i \neq j$, and the fact that Z_{n+1} on the boundary equals $\frac{1}{\sqrt{2}}\frac{\partial}{\partial\rho}$ + an (imaginary) tangential vector field. Then Lemma 6.20 and the previous calculations yield

<u>6.22 Proposition</u>. <u>Let</u> $\varphi = \sum_{j=1}^{n} \varphi_j \overline{\omega}^j$. <u>Then</u>

(6.23) $$\Box(\varphi) = (\overline{\partial}\vartheta + \vartheta\overline{\partial})(\varphi)$$

$$= \sum_{j=1}^{n+1} \left\{\left(-\frac{1}{2}\sum_{i=1}^{n}(Z_i\overline{Z}_i + \overline{Z}_i Z_i) - Z_{n+1}\overline{Z}_{n+1} - (2-n)iT\right)(\varphi_j)\right\}\overline{\omega}^j$$

$$+ ((h_{n+1}I_{n+1} + S^t_{n+1}) \circ (\overline{Z}_{n+1}\varphi))\overline{\omega} - (\overline{S}_{n+1} \circ (Z_{n+1}\varphi))\overline{\omega}$$

$$+ \varepsilon(Z, \overline{Z})(\varphi) + O(\rho)T(\varphi) + M_{n+1} + \varepsilon(\varphi).$$

Chapter 7. Local solution of the Dirichlet problem for $\Box$

Let $\omega_{n+1} = \sqrt{2}\,\partial\rho$ be the invariantly defined "complex normal" holomorphic covector near bM. Here ρ stands for geodesic distance from bM. Let $u \in C^\infty_{(0,1)}(M')$. Set

(7.1) $$u = u_t + \overline{\omega}_{n+1}\nu(u)$$

in a neighborhood of bM, where

(7.2) $$\nu(u) = \overline{\omega}_{n+1} \lrcorner\, u.$$

Clearly $[u_t]_{bM} \in C^\infty_{(0,1)}(bM)$, where $[\ \ ]_{bM}$ stands for the restrictions to bM. Let $u \in C^\infty_{(0,1)}(\overline{M})$ solve the following "Dirichlet" problem

(7.3) $$\Box(u) = f \quad \text{in } M$$

(7.4) $$[u_t]_{bM} = h,$$

(7.5) $$[\nu(u)]_{bM} = g.$$

The purpose of this chapter is to construct, locally, Green's operator $G: (f;h,g) \longrightarrow u$. The construction is quite technical, therefore we begin with a quick sketch of the main idea behind it (and is not intended to be precise).

Let E be a fundamental solution of $\Box$, i.e., $v = E(f)$ solves $\Box(v) = f$. We set $w = u-v$, u given in (7.3). Then to solve (7.3), (7.4) and (7.5) we only need to solve for

(7.6) $$\Box(w) = 0 \quad \text{in } M,$$

(7.7) $$[w_t]_{bM} = h',$$

(7.8) $$[\nu(w)]_{bM} = g'.$$

In general such a w is not unique,* but for the sake of heuristical considerations we shall assume that it is. Then the operator P,

(7.9) $$P: (h', g') \longrightarrow w$$

is called the Poisson operator for $\square$. To construct P one constructs first a fundamental solution E. Next let $v \in C^{\infty}_{(0,1)}(bM) \oplus C^{\infty}(bM)$. Let δ denote the δ-function of bM, i.e.,

(7.10) $$\delta(f) = \int_{bM} f(bM)\, d(bM).$$

Then $v \otimes \delta$ can be considered as a (0,1)-form on M' with support in bM. Consequently

(7.11) $$\square(E(v \otimes \delta)) = 0 \quad \text{in } M.$$

Define the operator $E_b: C^{\infty}_{(0,1)}(bM) \oplus C^{\infty}(bM) \longrightarrow C^{\infty}_{(0,1)}(M')$ by

(7.12) $$E_b(v) = E(v \otimes \delta).$$

Then

(7.13) $$[E_b]_{bM}(v) = [E_b(v)]_{bM}$$

is an elliptic pseudo-differential operator on bM. Let $[E_b]^{-1}_{bM}$ denote its inverse. Then, according to (7.11) the operator P defined by

(7.14) $$P: v \longrightarrow E([E_b]^{-1}_{bM}(v) \otimes \delta)$$

is the Poisson operator. In the rest of this chapter we shall use these ideas to construct explicitly local versions of the operators E, G and P

---

*since $\square$ has zero-order terms.

The existence and properties of these operators are by now well known. The purpose of the following lengthy calculations is to obtain precise local expressions for these operators which will be essential for the construction of the approximate Neumann's operator in Chapters 9 and 10. We use the calculus of pseudo-differential operators and boundary layer potentials. As such our treatment is based on the approach developed in detail by A. Calderón [5], L. Hörmander [18], R. Seeley [31] and L. Boutet de Monvel [2].

From the known a priori estimate for the $\bar{\partial}$-Neumann problem it is clear that we have to keep track of the principal part of the operators involved and also of those terms whose degree of homogeneity is one less than that of the principal part. This makes sense as long as we work in fixed local coordinates. Our final results concerning regularity and existence of solutions will, of course, be stated invariantly.

We fix, once and for all, a boundary coordinate neighborhood $U \subset M'$ with coordinates (x, ρ), $x = (x_1, \ldots, x_{2n+1})$. U is identified with a subset of $\mathbb{R}^{2n+2}$ so that $U \cap bM = \{(x, \rho) \in U;\ \rho = 0\}$.

Let us now briefly recall the main facts about pseudo-differential operators. In view of the notation we have adopted it will be simpler if we describe matters in the setting of bM, i.e., locally in $\mathbb{R}^{2n+1}$. The corresponding statements for the neighborhoods of M require then only the notational addition of two variables.

By a "classical" pseudo-differential operator of order k, we mean an operator T whose symbol $p(x, \xi)$ belongs to the class $S^k = S^k_{1,0}$;

i.e., p is jointly C^∞ in x and ξ, has compact support in the x-variable and satisfies for all α and β

$$|(\partial_\xi)^\alpha (\partial_x)^\beta p(x,\xi)| \leq C_{\alpha,\beta}(1+|\xi|)^{k-|\alpha|}.$$

The operator T defined by p is given as

$$(Tf)(x) = \frac{1}{(2\pi)^{2n+1}} \int_{\mathbb{R}^{2n+1}} e^{i\langle \xi, x\rangle} p(x,\xi)\hat{f}(\xi)\,d\xi.$$

We shall sometimes write T as T_k so as to indicate its order. If T_k and T_ℓ are two such pseudo-differential operators (corresponding to symbols $p(x,\xi)$ and $q(x,\xi)$) then their product is a pseudo-differential operator of order $k+\ell$ whose symbol $p \circ q$ has an asymptotic development

$$\sum_\alpha \frac{1}{i^{|\alpha|}\alpha!} \partial_\xi^\alpha p(x,\xi)\,\partial_x^\alpha q(x,\xi), \text{ in the sense that}$$

$$p \circ q - \sum_{|\alpha|<N} \frac{1}{i^{|\alpha|}\alpha!} \partial_\xi^\alpha p(x,\xi)\,\partial_x^\alpha q(x,\xi) \in S^{k+\ell-N}$$

For a rapid and highly readable introduction to pseudo-differential operators see Nirenberg [26]. It will be clear from the context whether our operators operate on M' or on bM. γ will denote the symbol of $\frac{\partial}{i\partial\rho}$, i.e., $\gamma = \sigma(\frac{\partial}{i\partial\rho})$. A pseudo-differential operator T_k defined on M', whose symbol is independent of γ will be denoted by T'_k. Working in local coordinates we will not distinguish between forms and vector functions

We begin by computing the symbol, $\sigma(\square)$, of $\square$.

$$\sigma(Z_{n+1}\overline{Z}_{n+1}) = \sigma\left(\frac{1}{\sqrt{2}}(\frac{\partial}{\partial\rho} + iT) \circ \frac{1}{\sqrt{2}}(\frac{\partial}{\partial\rho} - iT)\right)$$

$$= \frac{1}{2}\sigma\left(\frac{\partial^2}{\partial\rho^2} - iT_\rho + T\circ T\right)$$

$$= -\frac{1}{2}(\gamma^2 + \tau^2) + \frac{1}{2}\tau_\rho + \frac{1}{2}i\sum_{j=1}^{2n+1}\tau_{\xi_j}\tau_{x_j},$$

where $\gamma = \sigma(\frac{\partial}{i\partial\rho})$, $\tau = \sigma(-iT)$, $\tau_\rho = \dfrac{\partial\tau(x,\rho;\xi)}{\partial\rho}, \ldots,$etc. and $T_\rho = [\frac{\partial}{\partial\rho}, T]$ $= \frac{\partial}{\partial\rho}\circ T - T\circ\frac{\partial}{\partial\rho}$. Also $(\xi,\gamma)\in\mathbb{R}^{2n+2}$, $\xi = (\xi_1,\ldots,\xi_{2n+1})$. Similarly

$$\sigma\left(-\frac{1}{2}\sum_{j=1}^{n}(Z_j\overline{Z}_j + \overline{Z}_jZ_j)\right)$$

$$= \sum_{j=1}^{n}|\sigma(Z_j)|^2 + \frac{1}{i}\,\mathrm{Re}\sum_{j=1}^{n}\sum_{k=1}^{2n+1}\frac{\partial\sigma(Z_j)}{\partial\xi_k}\,\frac{\partial\overline{\sigma(Z_j)}}{\partial x_k}.$$

Therefore, using the notation of Proposition 6.22 we have

7.15. Lemma. *The symbol* $\sigma(\Box)$ *of* $\Box$ *is given by*

$$\sigma(\Box) = \left(\frac{1}{2}\gamma^2 + \frac{1}{2}\tau^2 + \sum_{j=1}^{n}|\sigma(Z_j)|^2\right)I_{n+1} \tag{7.16}$$

$$+ \left(-\frac{1}{2}\tau_\rho - \frac{1}{2}i\sum_{k=1}^{2n+1}\tau_{\xi_k}\tau_{x_k}\right.$$

$$\left.+\frac{1}{i}\,\mathrm{Re}\sum_{j=1}^{n}\sum_{k=1}^{2n+1}\frac{\partial\sigma(Z_j)}{\partial\xi_k}\,\frac{\partial\overline{\sigma(Z_j)}}{\partial x_k} + (2-n)\tau\right)I_{n+1}$$

$$+ \frac{1}{\sqrt{2}}(h_{n+1}I_{n+1} + S^t_{n+1})\sigma(\frac{\partial}{\partial\rho} - iT) - \frac{1}{\sqrt{2}}\overline{S}_{n+1}\sigma(\frac{\partial}{\partial\rho} + iT)$$

$$+ \epsilon(\sigma(Z),\sigma(\overline{Z})) + \sigma(M_{n+1}) + O(\rho)\sigma(T) + \sigma(T_0).$$

Here T_0 stands for multiplication by functions.

Next we compute the symbol of the fundamental solution E for $\Box$ in U. Let

$$(7.17)\qquad e(x,\rho;\xi,\gamma) \sim e_0(x,\rho;\xi,\gamma) + e_1(x,\rho;\xi,\gamma) + \cdots$$

denote the usual asymptotic expansion of the symbol e of E.

Let

$$(7.18)\qquad \sigma(\Box) = d_0 I_{n+1} + d_1 + d_2$$

be the decomposition of the symbol (7.16) in decreasing order of homogeneity (in ξ and γ), so that d_j has degree $2-j$. Write

$$(7.19)\qquad \Box = \Box_0 I_{n+1} + \Box_1 + \Box_2$$

where $\sigma(\Box_j) = d_j$.

Then we can obtain the first term of the asymptotic expansion as the inverse of the highest order term in (7.18), i e.,

$$(7.20)\qquad e_0(x,\rho;\xi,\gamma) = \left(\frac{1}{2}\gamma^2 + \frac{1}{2}\tau^2 + \sum_{j=1}^{n} |\sigma(Z_j)|^2\right)^{-1} I_{n+1}$$

$$= d_0^{-1} I_{n+1}.$$

Following the composition formula for symbols to the next order, we have

$$d_1 e_0 + d_0 e_1 + \frac{1}{i}(d_0)_\gamma (e_0)_\rho + \frac{1}{i}\sum_{k=1}^{2n+1} (d_0)_{\xi_k}(e_0)_{x_k} = 0.$$

Therefore

$$e_1 = -\frac{1}{d_0^2} d_1 - \frac{1}{d_0}\Big\{\frac{1}{i}(d_0)_\gamma \Big(\frac{1}{d_0}\Big)_\rho$$

$$+ \frac{1}{i}\sum_{k=1}^{2n+1} (d_0)_{\xi_k}\Big(\frac{1}{d_0}\Big)_{x_k}\Big\} I_{n+1}.$$

The terms in the bracket can be easily computed as follows.

$$(d_0)_\gamma \left(\frac{1}{d_0}\right)_\rho = \gamma\left(-\frac{1}{d_0^2}\right)(\tau\tau_\rho + \epsilon(\sigma(Z), \overline{\sigma(Z)})\, T_1'),$$

and

$$\sum_{k=1}^{2n+1} (d_0)_{\xi_k} \left(\frac{1}{d_0}\right)_{x_k}$$

$$= \sum_{k=1}^{2n+1} (\tau\tau_{\xi_k})\left(-\frac{1}{d_0^2}\right)(\tau\tau_{x_k}) + \epsilon(\sigma(Z), \overline{\sigma(Z)})\, \frac{1}{d_0^2}\, T_2',$$

where T_1' and T_2' have classical symbols of order one and two, respectively, which do not depend on γ. Collecting terms we obtain

$$\text{(7.21)} \qquad e_1 = -\frac{1}{d_0^2} d_1 + \frac{1}{i}\left(\frac{\gamma}{d_0^3}\tau\tau_\rho + \sum_{k=1}^{2n+1} \frac{1}{d_0^3}\tau_{\xi_k}\tau_{x_k}\right) I_{n+1}$$

$$+ \epsilon(\sigma(Z), \overline{\sigma(Z)})\, (T_2' + \gamma T_1')\frac{1}{d_0^3}.$$

Recall that

$$\text{(7.22)} \qquad d_1 = \left(-\frac{1}{2}\tau_\rho - \frac{1}{2} i \sum_{k=1}^{2n+1} \tau_{\xi_k}\tau_{x_k}\right.$$

$$\left. + \frac{1}{i}\operatorname{Re} \sum_{j=1}^{n} \sum_{k=1}^{2n+1} \frac{\partial\sigma(Z_j)}{\partial\xi_k}\, \frac{\overline{\partial\sigma(Z_j)}}{\partial x_k} + (2-n)\tau\right) I_{n+1}$$

$$+ \frac{1}{\sqrt{2}}(h_{n+1} I_{n+1} + S_{n+1}^t)\,\sigma\left(\frac{\partial}{\partial\rho} - iT\right) - \frac{1}{\sqrt{2}}\bar{S}_{n+1}\,\sigma\left(\frac{\partial}{\partial\rho} + iT\right)$$

$$+ \epsilon(\sigma(Z), \overline{\sigma(Z)}) + \sigma(M_{n+1}) + O(\rho)\,\sigma(T) + \sigma(T_0).$$

Following the heuristic discussion culminating in (7.14) we shall apply E to forms supported on bM, i.e., on $\rho = 0$. Therefore we integrate out the normal component γ of the cotangent bundle, i.e.,

(7.23) $$\frac{1}{2\pi}\int_{-\infty}^{\infty} e(x,\rho;\xi,\gamma)e^{i\rho\gamma}d\gamma$$

is the symbol of E *operating on* $\mathbb{R}^{2n+1}$. *These considerations suggest* that we separate the terms in d_1 that depend on γ from the terms which are independent from γ. To this end we set

(7.24) $$d_{11} = \Big\{-\frac{1}{2}\tau_\rho - \frac{1}{2}i\sum_{k=1}^{2n+1}\tau_{\xi_k}\tau_{x_k} + (2-n)\tau$$

$$+\frac{1}{i}\operatorname{Re}\sum_{j=1}^{n}\sum_{k=1}^{2n+1}\frac{\partial\sigma(Z_j)}{\partial\xi_k}\overline{\frac{\partial\sigma(Z_j)}{\partial x_k}}\Big\}I_{n+1}$$

$$+\frac{1}{\sqrt{2}}(h_{n+1}I_{n+1}+S^t_{n+1}+\overline{S}_{n+1})\tau$$

$$+\sigma(M_{n+1}) + O(\rho)\sigma(T),$$

and

(7.25) $$d_{12} = \frac{1}{\sqrt{2}}i(h_{n+1}I_{n+1}+S^t_{n+1}-\overline{S}_{n+1}).$$

Then

(7.26) $$\sigma(\Box) = d_0 I_{n+1} + d_{11} + d_{12}\gamma$$

$$+\epsilon(\sigma(Z),\overline{\sigma(Z)})$$

$$+ \text{multiplication by functions.}$$

We collect these calculations in the following form.

7.27. Lemma. Let E be a fundamental solution of $\Box$ in a neighborhood U of the origin in $\mathbb{R}^{2n+2}$, with coordinates (x,ρ), $x = (x_1,\ldots,x_{2n+1})$, i.e., $\Box E(f) = f$ in U if $f\in C^\infty_{(0,1)}(\mathbb{R}^{2n+2})$ with support in U. Then E is a pseudo-differential operator with symbol $e(x, \rho; \xi, \gamma)$, where $\gamma = \sigma(\frac{\partial}{i\partial\rho})$, and if

(7.28) $$e \approx e_0 + e_1 + \dots$$

stands for the usual asymptotic expansion, then

(7.29) $$e_0 = d_0^{-1} I_{n+1},$$

and

(7.30) $$e_1 = -\frac{1}{d_0^2} d_{11} - \frac{\gamma}{d_0^2} d_{12}$$

$$+ \frac{1}{i}\left(\frac{\gamma}{d_0^3} \tau\tau_\rho + \sum_{k=1}^{2n+1} \frac{1}{d_0^3} \tau^2 \tau_{\xi_k} \tau_{x_k}\right) I_{n+1}$$

$$+ \epsilon(\sigma(Z), \overline{\sigma(Z)}) (T_2' + \gamma T_1') d_0^{-3}.$$

Here d_{11} and d_{12} are given by (7.24) and (7.25), respectively.

Remark. Lemma 7.27 determines the first two terms of the asymptotic expansion of the symbol of the operator E. This is done under the assumption that such an operator exists. With a little extra work one can also prove the existence of such E. We sketch the idea. First by following out the complete asymptotic formula for the composition of two symbols we can construct a pseudo-differential operator E^*, so that $\Box E^* = I + S$, where S is an operator whose kernel is (jointly) C^∞. By restricting consideration to a sufficiently small neighborhood, the norm of S is then less than one, $(I + S)^{-1}$ exists and thus the operator $E = E^*(I + S)^{-1}$, satisfies $\Box E = I$. Since $\Box(E-E^*) = -S$, the local regularity of the operator

$\square$ insures that $E = E^* +$ an operator with a C^∞ kernel.

We now come to the operators of Poisson type.

7.31 Definition. $p(x, \rho; \xi)$ is a symbol of Poisson type of order j, j an integer, if it satisfies the following conditions.

(i) $p(x, \rho; \xi) \in C^\infty(\overline{\mathbb{R}^{2n+2}_+} \times \mathbb{R}^{2n+1})$, $(x, \rho) \in \overline{\mathbb{R}^{2n+2}_+}$, $\xi \in \mathbb{R}^{2n+1}$,

(ii) $p(x, \rho; \xi)$ has compact support in the (x, ρ) variables (in $\overline{\mathbb{R}^{2n+2}_+}$),

(iii)
$$\left|\rho^\delta \left(\frac{\partial}{\partial\rho}\right)^\gamma \left(\frac{\partial}{\partial x}\right)^\beta \left(\frac{\partial}{\partial\xi}\right)^\alpha p(x, \rho; \xi)\right|$$
$$\leq C_{\alpha,\beta,\gamma,\delta} (1+|\xi|)^{j-|\alpha|+\gamma-\delta}$$

$$\left(\frac{\partial}{\partial x}\right)^\beta = \left(\frac{\partial}{\partial x_1}\right)^{\beta_1} \cdots \left(\frac{\partial}{\partial x_{2n+1}}\right)^{\beta_{2n+1}}, \quad \text{etc.}$$

We note that p can be a scalar function or a matrix function. We also note, that as a result of this definition, $\rho^\delta\left(\frac{\partial}{\partial\rho}\right)^\gamma p(x, \rho; \xi)$ is, for each fixed ρ, a symbol of the standard class $S^{j+\gamma-\delta}_{1,0}$, and is so uniformly.

7.32 Definition. Let $p(x, \rho; \xi)$ be a symbol of Poisson type of order j. Then the mapping $P = f \longrightarrow F$, of a function f in $\mathbb{R}^{2n+1}$ to a function F in $\mathbb{R}^{2n+2}_+$, given by

$$F(x, \rho) = (2\pi)^{-2n-1} \int_{\mathbb{R}^{2n+1}} p(x, \rho; \xi) \hat{f}(\xi) e^{i\langle x, \xi\rangle} d\xi, \tag{7.33}$$

assuming the integral makes sense, is called an operator of Poisson type of order j.

In part III we shall study the behavior of operators of Poisson type of order zero on L^p- and Lipschitz spaces. This suffices to yield L^p- and Lipschitz estimates for operators of Poisson type of arbitrary order, since differentiating them on the left or multiplying them on the right by elements of $S^k_{1,0}(\mathbb{R}^{2n+1})$ again leads to symbols of Poisson type.

The next point is the result that E acting on forms on $\mathbb{R}^{2n+1}$ is an operator of Poisson type. We shall prove it is and we shall compute the relevant part of its symbol. This is accomplished in a series of lemmas.

Let $\psi(\xi,\gamma) = 1$ when $\gamma^2 + |\xi|^2 > 4$ and vanish in $\gamma^2 + |\xi|^2 < 1$, $\psi \in C^\infty(\mathbb{R}^{2n+2})$, and let $\varphi \in C_0^\infty(\mathbb{R}^{2n+2})$, $\varphi \equiv 1$ on V, $V \subset\subset U$. Suppose $f \in C^\infty_{(0,1)}(\mathbb{R}^{2n+1}) \oplus C^\infty(\mathbb{R}^{2n+1})$ with support in V. Let E_0 denote the pseudo-differential operator induced by the symbol $\varphi(x,\rho)\, e_0(x,\rho;\xi,\gamma)\,\psi(\xi,\gamma)$. We set

(7.34) $$E_{b,0}(f) = E_0(f \otimes \delta)$$

where

(7.35) $$\delta(g) = \int_{\mathbb{R}^{2n+1}} g(x,0)\,dx\ , \quad g \in C_0^\infty(U).$$

<u>7.36. Lemma.</u> <u>Let</u> $E_{b,0}: C^\infty_{(0,1)}(\mathbb{R}^{2n+1}) \oplus C^\infty(\mathbb{R}^{2n+1}) \longrightarrow C^\infty_{(0,1)}(\mathbb{R}^{2n+2})$ <u>be defined by</u> (7.34). <u>Then</u> $E_{b,0}$ <u>is an operator of Poisson type of order</u> -1. <u>More precisely</u>

(7.37) $$E_{b,0}(f)(x,\rho)$$
$$= (2\pi)^{-2n-1} \int_{\mathbb{R}^{2n+1}} e^{i\langle x,\xi\rangle} (\varphi(x,\rho)\, e_{b,0}(x,\rho;\xi)\,\psi(\xi)$$
$$+ r_b(x,\rho;\xi))\,\hat{f}(\xi)\,d\xi\ ,$$

$f \in C^{\infty}_{(0,1)}(\mathbb{R}^{2n+1}) \oplus C^{\infty}(\mathbb{R}^{2n+1})$ with support in V, where

(7.38) (i) $$e_{b,0}(x,\rho;\xi) = \frac{1}{2\pi}\int_{-\infty}^{\infty} e^{i\rho\gamma}\left(\frac{1}{2}\gamma^2 + \frac{1}{2}\tau^2 + \sum_{j=1}^{n} |\sigma(Z_j)|^2\right)^{-1} d\gamma$$

$$= \frac{e^{-\rho\Delta(x,\rho;\xi)}}{\Delta(x,\rho;\xi)},$$

$$\Delta(x,\rho;\xi) = \left(\tau^2 + 2\sum_{j=1}^{n} |\sigma(Z_j)|^2\right)^{1/2},$$

(ii) $\psi(\xi) \in C^{\infty}(\mathbb{R}^{2n+1})$, $\psi \equiv 1$ if $|\xi| > 2$ and vanishes if $|\xi| < 1$, and

(iii) $r_{b,-\infty}(x,\rho;\xi)$ is a symbol of Poisson type of order $-\infty$.

The gist of the proof of this lemma is the simple identity (i). The detailed proof of a more general form of this lemma can be found in Hörmander [18], Theorem 2.14.

To simplify matters we use the Taylor expansion of $\Delta(x,\rho;\xi)$ with respect to ρ about $\rho = 0$. This easily yields, (with $\Delta(x,0,\xi) = \Delta$),

7.39. Lemma.

$$\frac{e^{-\rho\Delta(x,\rho;\xi)}}{\Delta(x,\rho;\xi)} = \frac{e^{-\rho\Delta}}{\Delta}$$

$$+ \rho\left(\frac{e^{-\rho\Delta}}{\Delta^2}(-\rho)\tau\tau_\rho - \frac{e^{-\rho\Delta}}{\Delta^3}\tau\tau_\rho\right)$$

$$+ r_{b,-3}(x,\rho;\xi) + r_{b,-2}(x,\rho;\xi) \in (\sigma(Z), \overline{\sigma(Z)})$$

where $\varphi(x,\rho)\, r_{b,-2}(x,\rho;\xi)\,\psi(\xi)$ and $\varphi(x,\rho)\, r_{b,-3}(x,\rho;\xi)\,\psi(\xi)$ are symbols

<u>of Poisson type of order</u> -2 <u>and</u> -3, <u>respectively</u>, Δ, τ <u>and</u> τ_ρ <u>are evaluated at</u> $\rho = 0$.

In a similar manner let E_1 be induced by the symbol $\varphi(x, \rho) e_1(x, \rho; \xi', \gamma) \psi(\xi, \gamma)$. Define

$$(7.40) \qquad E_{b,1}(f) = E_1(f \otimes \delta),$$

$f \in C^\infty_{(0,1)}(\mathbb{R}^{n+1}) \oplus C^\infty(\mathbb{R}^{n+1})$ with support in $V \subset\subset U$. Then $E_{b,1}$ is induced by the symbol $\varphi(x, \rho) e_{b,1}(x, \rho; \xi) \psi(\xi') + r_{b,-\infty}(x, \rho; \xi)$ of Poisson type, where

$$(7.41) \qquad e_{b,1}(x, \rho; \xi) = \frac{1}{2\pi} \int_{-\infty}^{\infty} e^{i\rho\gamma} e_1(x, \rho; \xi, \gamma)\, d\gamma$$

and $r'_{b,-\infty}(x, \rho; \xi)$ is a symbol of Poisson type of order $-\infty$. To calculate $e_{b,1}$ more precisely we need the following evaluation of integrals. (Note that $\Delta^2 = \tau^2 + 2\Sigma|\sigma(Z_j)|^2$, and $d_0 = \frac{1}{2}(\gamma^2 + \Delta^2)$.)

$$\frac{1}{2\pi} \int_{-\infty}^{\infty} \frac{e^{i\rho\gamma} d\gamma}{d_0^2} = \frac{\rho e^{-\rho\Delta}}{\Delta^2} + \frac{e^{-\rho\Delta}}{\Delta^3},$$

$$\frac{1}{2\pi} \int_{-\infty}^{\infty} \frac{\gamma e^{-\rho\gamma} d\gamma}{d_0^2} = \frac{i\rho e^{-\rho\Delta}}{\Delta},$$

$$\frac{1}{2\pi i} \int_{-\infty}^{\infty} \frac{e^{i\rho\gamma} d\gamma}{d_0^3} = \frac{\rho^2 e^{-\rho\Delta}}{2i\Delta^3} + \frac{3\rho e^{-\rho\Delta}}{2i\Delta^4} + \frac{3e^{-\rho\Delta}}{2i\Delta^5},$$

$$\frac{1}{2\pi i} \int_{-\infty}^{\infty} \frac{\gamma e^{i\rho\gamma} d\gamma}{d_0^3} = \frac{\rho e^{-\rho\Delta}}{2\Delta^3} + \frac{\rho^2 e^{-\rho\Delta}}{2\Delta^2}.$$

Thus we have the following result.

7.42. Lemma. $E_{b,1}$ is induced by the following symbol of Poisson type

$$\varphi(x) e_{b,1}(x, \rho; \xi) \psi(\xi) + r_{b,-3}(x, \rho; \xi),$$

where

$$e'_{b,1}(x, \rho; \xi) = -\left(\frac{\rho e^{-\rho\Delta}}{\Delta^2} + \frac{e^{-\rho\Delta}}{\Delta^3}\right) d_{11}$$

$$- i \frac{\rho e^{-\rho\Delta}}{\Delta} d_{12}$$

$$+ \left(\frac{\rho e^{-\rho\Delta}}{2\Delta^3} + \frac{\rho^2 e^{-\rho\Delta}}{2\Delta^2}\right) \tau \tau_\rho I_{n+1}$$

$$+ \left(\frac{\rho^2 e^{-\rho\Delta}}{2i\Delta} + \frac{3\rho e^{-\rho\Delta}}{2i\Delta^2} + \frac{3 e^{-\rho\Delta}}{2i\Delta^3}\right)$$

$$\cdot \left(1 - \frac{2\sum_{j=1}^{n} |\sigma(Z_j)|^2}{\Delta^2}\right) \sum_{k=1}^{2n+1} \tau_{\xi_k} \tau_{x_k} I_{n+1} .$$

The coefficients of τ in $e_{b,1}$ are evaluated at $\rho = 0$ and $r_{b,-3}(x, \rho; \xi)$ is a symbol of Poisson type of order -3.

We remark that $E - (E_0 + E_1)$ applied to forms on $\rho = 0$ induces an operator of Poisson type of order -3, hence its precise form is irrelevant.

Finally, by summing the relevant terms in Lemmas 7.36, 7.39 and setting $\rho = 0$ in the coefficients of τ and $\sigma(Z_j)$, $j=1,\ldots,n$, we obtain the necessary expression for the symbol e_b of the operator E_b as follows

7.43. Proposition. Let $f \in C^{\infty}_{(0,1),0}(\mathbb{R}^{2n+1} \cap V) \oplus C^{\infty}_{0}(\mathbb{R}^{2n+1} \cap V)$. The operator E_b given by

$$E_b(f) = E(f \otimes \delta) \tag{7.44}$$

is induced by a symbol $e_b(x, \rho; \xi)$ of Poisson type given by

$$e_b(x, \rho; \xi) = \varphi(x, \rho)(e_{b,0}(x, \rho; \xi) + e_{b,1}(x, \rho; \xi))\psi(\xi) \tag{7.45}$$
$$+ r_{b,-2}(x, \rho; \xi)\, \varepsilon(\sigma(Z), \overline{\sigma(Z)}) + r_{b,-3}(x, \rho; \xi),$$

where

(i) $e_{b,0}(x, \rho; \xi) = \dfrac{e^{-\rho\Delta}}{\Delta} I_{n+1}$,

(ii) $e_{b,1}(x, \rho; \xi)$

$$= -\left(\frac{\rho e^{-\rho\Delta}}{\Delta^2} + \frac{e^{-\rho\Delta}}{\Delta^3}\right) d_{11} - i\frac{\rho e^{-\rho\Delta}}{\Delta} d_{12}$$

$$- \left(\frac{\rho e^{-\rho\Delta}}{2\Delta^3} + \frac{\rho^2 e^{-\rho\Delta}}{2\Delta^2}\right) \tau\, \tau_\rho I_{n+1}$$

$$+ \left(\frac{\rho^2 e^{-\rho\Delta}}{2i\,\Delta} + \frac{3\rho e^{-\rho\Delta}}{2i\,\Delta^2} + \frac{3 e^{-\rho\Delta}}{2i\,\Delta^3}\right)$$

$$\cdot \left(1 - \frac{2\sum_{j=1}^{n} |\sigma(Z_j)|^2}{\Delta^2}\right) \tau_{\xi_k} \tau_{x_k} I_{n+1},$$

(iii) $r_{b,-2}(x, \rho; \xi)$ and $r_{b,-3}(x, \rho; \xi)$ are symbols of Poisson type of order -2 and -3, respectively,

(iv) $\psi(\xi) \in C^{\infty}(\mathbb{R}^{2n+1})$, $\psi = 1$ if $|\xi| > 2$ and vanishes in $|\xi| < 1$, and $\varphi \in C_0^\infty(\mathbb{R}^{2n+2} \cap U)$, $\varphi \equiv 1$ on V, $V \subset\subset U$.

Furthermore

(7.46) $$\square \circ E_b(f) = 0 \ \text{in} \ V \cap \mathbb{R}_+^{2n+2}.$$

Let $[E_b]_0$ denote the restriction of E_b to $bM = \{x \in \mathbb{R}^{2n+2}; \rho = 0\}$, i.e.,

(7.47) $$[E_b]_0(f) = \lim_{\rho \to 0} E_b(f),$$

$f \in C_{(0,1)}^\infty(\mathbb{R}^{2n+1}) \oplus C_0^\infty(\mathbb{R}^{2n+1})$ with support in $V \cap \mathbb{R}^{2n+1}$. Then $[E_b]_0$ is an elliptic pseudo-differential operator with symbol

$$[e_b]_0 (x, \xi) = e_b(x, 0; \xi), \ \text{where}$$

(7.48) $$e_b(x, 0; \xi) \approx \Delta^{-1} I_{n+1} - \Delta^{-3} d_{11}$$

$$+ \frac{3}{2i} \Delta^{-3} \left(1 - \frac{2 \sum_{j=1}^n |\sigma(Z_j)|^2}{\Delta^2}\right) \left(\sum_{k=1}^{2n+1} \tau_{\xi_k} \tau_{x_k} \right) I_{n+1}$$

$$+ \sigma(T_{-3}) + \sigma(T_{-2}) \epsilon (\sigma(Z), \overline{\sigma(Z)}).$$

Let $V_0 = V \cap \mathbb{R}^{2n+1}$, where $\mathbb{R}^{2n+1} = \{x \in \mathbb{R}^{2n+2}; \rho = 0\}$. If V is sufficiently small, then $[E_b]_0 = \lim_{\rho \to 0} E_b$ has an inverse, $[E_b]_0^{-1}$, in V_0, i.e., suppose $f \in C_{(0,1)}^\infty(\mathbb{R}^{2n+1}) \oplus C^\infty(\mathbb{R}^{2n+1})$ with support in V_0. Then we have,

(7.49) Lemma

(i) $\lim_{\rho \to 0} E_b \circ [E_b]_0^{-1} (f) = f$ in V_0,

(ii) $[E_b]_0^{-1} \circ (\lim_{\rho \to 0} E_b(f)) = f,$

(iii) $[E_b]_0^{-1}$ <u>is a pseudo-differential operator with symbol</u> $[e_b]_0^{-1}(x, \xi) \in S^1_{1,0}$, <u>where</u>

(7.50)
$$[e_b]_0^{-1}(x,\xi) = \Delta I_{n+1} + \Delta^{-1} d_{11}$$
$$+ \frac{i}{2\Delta}\left(1 - \frac{2\sum_{j=1}^{n} |\sigma(Z_j)|^2}{\Delta^2}\right)\left(\sum_{k=1}^{2n+1} \tau_{\xi_k} \tau_{x_k}\right) I_{n+1}$$
$$+ \sigma(T_{-1}) + \epsilon(\sigma(Z), \overline{\sigma(Z)})\, \sigma(T_{-1}).$$

<u>Proof</u>. The result follows easily from the above by the classical pseudo-differential operator calculus and then by an additional argument, as in the Remark following Lemma 7.27.

We define the operator P,

(7.51)
$$P: C^{\infty}_{(0,1),0}(\mathbb{R}^{2n+1}) \oplus C^{\infty}_0(\mathbb{R}^{2n+1}) \longrightarrow C^{\infty}_{(0,1),0}(\mathbb{R}^{2n+2}_+)$$

by

(7.52)
$$P = E_b \circ [E_b]_0^{-1}.$$

Then a combination of (7.45) and (7.50) gives

(7.53)
$$P = P_0 + P_1$$

where P and P_j are induced by the symbols p and p_j of poisson type, and

(7.54) $p(x, \rho; \xi) = \varphi(x, \rho)(p_0(x, \rho; \xi) + p_1(x, \rho; \xi))\, \psi(\xi)$, where

(7.55)
$$p_0(x, \rho; \xi) = e^{-\rho\Delta} I_{n+1}$$

$$(7.56)\qquad p_1(x,\rho;\xi) = -\rho e^{-\rho\Delta}\frac{1}{\Delta}d_{11} - i\rho e^{-\rho\Delta}d_{12}$$

$$-\left(\rho e^{-\rho\Delta}\frac{1}{2\Delta^2} + \rho^2 e^{-\rho\Delta}\frac{1}{2\Delta}\right)\tau\tau_\rho I_{n+1}$$

$$+\frac{1}{2i}\left(\rho^2 e^{-\rho\Delta} + \rho e^{-\rho\Delta}\frac{1}{\Delta}\right)\sum_{k=1}^{2n+1}\tau_{\xi_k}\tau_{x_k} I_{n+1}$$

$$+ r_{b,-2}(x,\rho;\xi) \in (\sigma(Z), \overline{\sigma(Z)}) + r'_{b,-2}(x,\rho;\xi).$$

Also

(i) $r_{b,-2}(x,\rho;\xi)$ and $r'_{b,-2}(x,\rho;\xi)$ are symbols of Poisson type of order -2,

(ii) $\psi \in C^\infty(\mathbb{R}^{2n+1})$, $\psi \equiv 1$ when $|\xi| > 2$ and $\psi \equiv 0$ in $|\xi| < 1$,

(iii) $\varphi \in C_0^\infty(U)$, $\varphi \equiv 1$ on V, $V \subset\subset U$.

(7 46) and Lemma 7.49 can now be put in the following form.

<u>7.57. Proposition.</u> <u>Let</u> $g \in C^\infty_{(0,1),0}(\mathbb{R}^{2n+1} \cap V) + C_0^\infty(\mathbb{R}^{2n+1} \cap V)$. <u>Then</u>

(i) $\square(P(g)) = 0$ <u>in</u> $\mathbb{R}_+^{2n+2} \cap V$,

(ii) $\lim_{\rho\to 0} P(g) = g$ <u>in</u> $\mathbb{R}^{2n+1} \cap V$.

With a slight abuse of language P is a "local Poisson operator."

Before we state the final result of this chapter we shall need an operator which extends functions given in the closed upper-half space $\overline{\mathbb{R}}_+^{2n+2}$ to the whole of $\mathbb{R}^{2n+2}$.

Suppose $f \in C^\infty(\overline{\mathbb{R}}_+^{2n+2} \cap U)$ and has compact support in U. Extend f to $\tilde{f}$ in $\overline{\mathbb{R}}_+^{2n+2}$ in the obvious way (i.e., f=0 outside U). Next, in R_-^{2n+2} set

$$\tilde{f}(x,\rho) = \int_1^\infty \tilde{f}(x,\rho(1-2\lambda))\,\psi(\lambda)\,d\lambda, \qquad \rho < 0 \tag{7.58}$$

where ψ is rapidly decreasing as $\lambda \longrightarrow \infty$ and

$$\int_1^\infty \psi(\lambda)\,d\lambda = 1, \qquad \int_1^\infty \lambda^k \psi(\lambda)\,d\lambda = 0, \quad k=1,2,\ldots . \tag{7.59}$$

Then $\tilde{f} \in C_0^\infty(\mathbb{R}^{2n+2})$, and clearly $\tilde{f} = f$ in $\mathbb{R}_+^{2n+1} \cap U$.

(7.60) Definition. We shall say that $\tilde{f}$ extends f. For further details about this extension operator, see [33], Chapter VI, §3.

The following additional remarks are useful in the applications:

(i) The operator $f \longrightarrow \tilde{f}$ clearly commutes with the differentiations $\frac{\partial}{\partial x_j}$, $j=1,\ldots,2n+1$.

(ii) $\frac{\partial(\tilde{f})}{\partial\rho} = (\frac{\partial f}{\partial \rho})^\sim$, where the second extension is of the same type as (7.58), except $\psi(\lambda)$ is replaced by $\psi_1(\lambda) = (1-2\lambda)\psi(\lambda)$. Now ψ_1 satisfies the condition (7.59) also, and since any ψ satisfying (7.59) will do, we shall not bother to introduce notation to distinguish these extensions.

(iii) If T_k is any pseudo-differential operator acting on functions defined in $\mathbb{R}^{2n+2}$, we extend it to act on functions which are in $C_0^\infty(\overline{\mathbb{R}}_+^{2n+1} \cap U)$, by setting

$$\tilde{T}_k(f) = T_k(\tilde{f}). \tag{7.61}$$

The particular extension operator (7.58) (i e., the particular ψ) used will not be relevant, and will not be kept track of.

The main result of this chapter concerning a local solution of the inhomogeneous Dirichlet problem follows.

Let U be a boundary coordinate neighborhood in M' with coordinates (x, ρ), $x = (x_1, \ldots, x_{2n+1})$, such that $U \cap bM = \{(x, \rho) \in \mathbb{R}^{2n+2}; \rho = 0\}$ if we identify U with a subset of $\mathbb{R}^{2n+2}$. By choosing a sufficiently small boundary neighborhood V, (i.e., $[E_b]_0$ has an inverse in $V_0 = V \cap \{\rho = 0\}$, $V \subset\subset U \subset \mathbb{R}^{2n+2}$) we have the following solution to the inhomogeneous Dirichlet problem for $\square$ in V.

7.62. Theorem (i) Let $f \in C^{\infty}_{(0,1),0}(\overline{\mathbb{R}^{2n+2}_{+}} \cap V)$ and $g \in C^{\infty}_{(0,1),0}(\mathbb{R}^{2n+1} \cap V) \oplus C^{\infty}_{0}(\mathbb{R}^{2n+1} \cap V)$. Let E be the local fundamental solution for $\square$ given in Lemma 7.27 and let P be the operator of Poisson type defined by (7.52). We set

(7.63) $$G(f) = \tilde{E}(f) - P([\tilde{E}(f)]_0)$$

where the passage from E to $\tilde{E}$ is as in (7.61).

Then

(7.64) $$\square(u) = f \text{ in } V \cap \mathbb{R}^{2n+2}_{+},$$

(7.65) $$[u]_0 = g \text{ in } V \cap \mathbb{R}^{2n+1},$$

where we set

(7.66) $$u = G(f) + P(g).$$

The above result may be viewed as the "existence" part of the solution of the Dirichlet problem, in the local version we are working in. The theorem below leads to the desired "regularity" results. For this we choose another neighborhood V_1, so that $V_1 \subset\subset V$.

7.67. Theorem. Suppose $u \in C^{\infty}_{(0,1),0}(\overline{\mathbb{R}}^{2n+2}_{+} \cap V_1)$, then in $\overline{\mathbb{R}}^{2n+1}_{+} \cap V_1$,

(7.68) $$u = G_1(\Box u) + P_1([u]_0)$$

where $G - G_1$ and $P - P_1$ are integral operators with C^{∞} kernels (on $\overline{\mathbb{R}}^{2n+2}_{+} \cap V_1 \times \overline{\mathbb{R}}^{2n+2}_{+} \cap V_1$ and $\overline{\mathbb{R}}^{2n+2}_{+} \cap V_1 \times \overline{\mathbb{R}}^{2n+1} \cap V_1$, respectively).

Proof. Let

(7.69) $\quad v = G(\Box u) + P([u]_0)$,

and $w = u - v$. Then in $\overline{\mathbb{R}}^{2n+2} \cap V$, $\Box w = 0$ and $[w]_0 = 0$, by the previous theorem. Next let $G(\overline{x}, \overline{y})$ denote the kernel of the operator G defined by (7.63). Here $\overline{x} = (x, \rho)$, $\overline{y} = (y, r)$. Following through the definitions of E, $\widetilde{E}$, and P we observe that $G(\overline{x}, \overline{y})$ is C^{∞} away from the diagonal. Moreover (7.64) implies that $\Box_{\overline{x}} G(\overline{x}, \overline{y}) = \delta_{\overline{y}}$, and (7.66) $[G(\overline{x}, \overline{y})]_{\rho=0} = 0$.

Next let U_0 be a smooth open domain in $\overline{\mathbb{R}}^{2n+2}_{+}$ so that $\overline{U}_0 \subset \overline{\mathbb{R}}^{2n+2}_{+} \cap V$, while $\overline{R}^{2n+2}_{+} \cap \overline{V}_1 / \mathbb{R}^{2n+1} \cap \overline{V}_1 \subset U_0$. We can then decompose the boundary of U_0 into two parts: $bU_0 = (bU_0)^0 \cup (bU_0)^+$. Here $(bU_0)^0$ lies on the hyperplane $\mathbb{R}^{2n+1}$, and $(bU_0)^+$ is the rest of the boundary (lying in $\mathbb{R}_+^{2n+2}$). What is crucial is that $(bU_0)^+$ is at a <u>positive</u> distance from V_1.

We apply Green's theorem to the form w and the region U_0 using the measure $dm(\bar{y})$ induced by our "Levi-metric" (see Chapter 5), for which $\square$ is self-adjoint.

Therefore, if $\bar{y} \in \mathbb{R}_+^{2n+2} \cap V_1$

$$\int_{U_0} (\square_{\bar{x}}\, G(\bar{x},\bar{y}))\, w(\bar{x})\, dm(\bar{x}) - \int_{U_0} \square_{\bar{x}} w(\bar{x})\, G(\bar{x},\bar{y})\, dm(\bar{x}) \tag{7.70}$$

$$= \int_{bU_0} A(\bar{x},\bar{y}) \cdot w(\bar{x})\, d\sigma(x) + \int_{bU_0} B(x,\bar{y}) \cdot (\nabla w)(\bar{x})\, d\sigma(\bar{x}) .$$

Here A is a combination with smooth coefficients of first derivatives of $G(\bar{x},\bar{y})$, while $B \cdot \nabla w$ is a linear combination of G with smooth coefficients and of first derivatives of w.

The left-hand side of (7.70) gives $w(\bar{y})$. Since both $w(\bar{x})$ and $G(\bar{x},\bar{y})$ vanish when $\bar{x} \in (bU_0)^0$ the right-hand side (7.70) needs to be evaluated on $(bU_0)^+$ only. But $w = u - v$ and $u = 0$ there, hence there $w(\bar{x}) = -v(\bar{x})$. Now when $\bar{x} \in (bU_0)^+$ and $\bar{y} \in V_1$, then their distance is strictly positive, so the functions A and B are smooth. Moreover for $\bar{x} \in (bU_0)^+$, by the formulas (7.69) $v(\bar{x})$ and $(\nabla \cdot v)(\bar{x})$ can be expressed in terms of $\square u$ on $[u]_0$, (which are supported in V_1) with C^∞ kernels. Integrating out in (7.70) shows that w is expressed in terms of integral operators with C^∞ kernels acting on $\square u$ and $[u]_0$. This completes the proof of the theorem.

Chapter 8. Reduction to the boundary

Here we compute the $\bar{\partial}$-Neumann boundary value of the Poisson operator P. Recall that P is defined by (7.52) on $C^{\infty}_{(0,1),0}(\mathbb{R}^{2n+1}) \oplus C^{\infty}_{0}(\mathbb{R}^{2n+1})$. We change the meaning of P slightly keeping the same notation, and hopefully not introducing ambiguity.

8.1 Definition. We define the operator

$$P: C^{\infty}_{(0,1),0}(\mathbb{R}^{2n+1}) \longrightarrow C^{\infty}_{(0,1)}(\mathbb{R}^{2n+2}_{+})$$

by

$$P(g) = P(g \oplus 0), \quad g \in C^{\infty}_{(0,1),0}(\mathbb{R}^{2n+1}).$$

Now let $u \in C^{\infty}_{(0,1),0}(\overline{\mathbb{R}^{2n+2}_{+}} \cap U)$ be given by

(8.2) $$u = \sum_{j=1}^{n+1} u_j \bar{\omega}^j = u_t + (\nu u)\bar{\omega}_{n+1}.$$

The first $\bar{\partial}$-Neumann boundary condition is

(8.3) $$[\nu u]_0 = 0.$$

According to (6.16), (and (7.1)) the second condition is the vanishing of

(8.4) $$[\nu \bar{\partial}(u)]_0 = [(\bar{Z}_{n+1} I_n + \bar{S}_{n+1}) u_t]_0.$$

We therefore introduce the short-hand of the operator $B_{\bar{\partial}}$ as follows.

(8.5) $$B_{\bar{\partial}}(u) = [(\bar{Z}_{n+1} I_n + \bar{S}_{n+1}) u_t]_0 = [\nu \bar{\partial}(u)]_0$$

for $u \in C^{\infty}_{(0,1)}$ with $[\nu u]_0 = 0$.

Here, with a slight abuse of notation, S_{n+1} stands for the $n \times n$

matrix which is obtained by dropping the n+1-st row and the n+1-st column of S_{n+1}, which are zero, anyway (see (6.12)).

8.6. Definition. We define the pseudo-differential operator $\overset{+}{\Box}; C^\infty_{(0,1),0}(\mathbb{R}^{2n+1}) \longrightarrow C^\infty_{(0,1),0}(\mathbb{R}^{2n+1})$ by

$$\overset{+}{\Box}(h) = [\nu\bar{\partial}P(h)]_0 = B_{\bar{\partial}}\, P(h),\ h \in C^\infty_{(0,1),0}(\mathbb{R}^{2n+1}). \tag{8.7}$$

8.8. Theorem. $\overset{+}{\Box}$ is induced by the following symbol

$$\sigma(\overset{+}{\Box}) = \varphi(x,0)\overset{+}{\Box}_s\, \psi(\xi) + [\overline{Z}_{n+1}\,\varphi]_0 \tag{8.9}$$
$$+ \varphi(x,0)(1 - \tfrac{T}{\Delta})\,\psi(\xi)\,\sigma(T_0)$$
$$+ \epsilon(\sigma(Z), \overline{\sigma(Z)})\,\sigma(T_{-1}) + \sigma(T_{-1}),$$

where

$$\overset{+}{\Box}_s = \frac{1}{\sqrt{2}}(\tau - \Delta)I_n - \frac{1}{\sqrt{2}}(2-n)\frac{\tau}{\Delta}I_n \tag{8.10}$$
$$- \frac{1}{i\sqrt{2}\,\Delta}\,\mathrm{Re}\sum_{j=1}^{n}\sum_{k=1}^{2n+1}\frac{\partial\sigma(Z_j)}{\partial\xi_k}\,\frac{\partial\overline{\sigma(Z_j)}}{\partial x_k}\,I_n,$$

and φ and ψ are defined in Proposition 7.57(ii), (iii).

Proof. Using the symbol of P given in (7.54), (7.55) and (7.56) we shall compute $\sigma(\overset{+}{\Box})$. To simplify matters first we note that the relevant symbols are diagonal with the exception of S_{n+1}. Namely, according to (7.55) p_0 is diagonal. So is p_1 by (7.56) with the exception of

$$-\rho e^{-\rho\Delta}\frac{1}{\Delta}d_{11} - i\rho e^{-\rho\Delta}d_{12}$$

and the error terms

$$r_{b,-2}(x,\rho;\xi)\,\epsilon(\sigma(Z),\overline{\sigma(Z)}) + r'_{b,-2}(x,\rho;\xi).$$

These error terms will be incorporated in the error terms of $\square^+$, therefore we shall neglect them. d_{11} and d_{12} are diagonal, except for terms containing S_{n+1}, which, according to our convention is considered to be an $n \times n$ matrix, and for the terms $\sigma(M_{n+1})$ and $O(\rho)\sigma(T)$. $\sigma(M_{n+1})$ does not even enter the calculation and $O(\rho)\sigma(T)$ contributes only to the error terms $r'_{b,-2}(x,\rho;\xi)$, thus negligible. Now the calculation of (8.9) is straightforward. Namely, if we leave off φ and ψ, we obtain

$$\sigma\left[\left(\frac{1}{\sqrt{2}}\frac{dP}{d\rho} - \frac{1}{\sqrt{2}}\, iTP + \overline{S}_{n+1}P\right)\right]_0$$

$$\approx \left(\frac{1}{\sqrt{2}}\tau - \frac{1}{\sqrt{2}}\Delta\right) + \frac{1}{\sqrt{2}}\,\frac{1}{2i\Delta}\left(\sum_{k=1}^{2n+1}\tau_{\xi_k}\tau_{x_k}\right)I_n$$

$$- \frac{1}{\sqrt{2}}\,\frac{\tau}{\Delta}\,\frac{\tau_\rho}{2\Delta}\,I_n - \frac{1}{\sqrt{2}}\,\frac{1}{\Delta}\,d_{11} - \frac{1}{\sqrt{2}}\,id_{12} + \overline{S}_{n+1}$$

$$+ \epsilon(\sigma(Z),\overline{\sigma(Z)})\,T_{-1} + T_{-1}$$

$$= \frac{1}{\sqrt{2}}(\tau - \Delta)I_n + \frac{1}{2\sqrt{2}\,i\Delta}\sum_{k=1}^{2n+1}\tau_{\xi_k}\tau_{x_k}I_n$$

$$- \frac{1}{\sqrt{2}}\,\frac{\tau}{\Delta}\,\frac{\tau_\rho}{2\Delta}\,I_n$$

$$- \frac{1}{\sqrt{2}\,\Delta}\left\{-\frac{1}{2}\tau_\rho - \frac{1}{2}i\sum_{k=1}^{2n+1}\tau_{\xi_k}\tau_{x_k} + (2-n)\tau\right.$$

$$\left. + \frac{1}{i}\,\mathrm{Re}\sum_{j=1}^{n}\sum_{k=1}^{2n+1}\frac{\partial\sigma(Z_j)}{\partial\xi_k}\,\frac{\partial\overline{\sigma(Z_j)}}{\partial x_k}\right\}I_n$$

$$- \frac{1}{2\Delta}(h_{n+1} I_n + S^t_{n+1} + \overline{S}_{n+1})\tau$$

$$+ \frac{1}{2}(h_{n+1} I_n + S^t_{n+1} - \overline{S}_{n+1}) + \overline{S}_{n+1}$$

$$+ \varepsilon(\sigma(Z), \overline{\sigma(Z)})\sigma(T_{-1}) + \sigma(T_{-1}).$$

Simplifying we obtain

$$\sigma(\Box^+) \approx \frac{1}{\sqrt{2}}(\tau - \Delta)I_n - \frac{1}{\sqrt{2}}(2-n)\frac{\tau}{\Delta}I_n$$

$$- \frac{1}{i\sqrt{2}\,\Delta} \operatorname{Re} \sum_{j=1}^{n} \sum_{k=1}^{2n+1} \frac{\partial \sigma(Z_j)}{\partial \xi_k} \frac{\partial \overline{\sigma(Z_j)}}{\partial x_k} I_n$$

$$+ \frac{\tau_\rho}{2\sqrt{2}\,\Delta}(1 - \frac{\tau}{\Delta})I_n + \frac{1}{2}\overline{S}_{n+1}(1 - \frac{\tau}{\Delta})$$

$$+ \frac{1}{2}(h_{n+1} I_n + S^t_{n+1})(1 - \frac{\tau}{\Delta})$$

$$+ \varepsilon(\sigma(Z), \overline{\sigma(Z)})\sigma(T_{-1}) + \sigma(T_{-1}).$$

This proves Theorem 8.8.

8.11. Definition. We define the pseudo-differential operator $\Box^- : C^\infty_{(0,1),0}(\mathbb{R}^{2n+1}) \longrightarrow C^\infty_{(0,1),0}(\mathbb{R}^{2n+1})$ by giving its symbol, $\sigma(\Box^-)$, as follows

(8.12) $$\sigma(\Box^-) = \varphi(x,0)\,\Box^-_s\,\psi(\xi),$$

where φ and ψ were defined in Proposition 7.57(ii),(iii) and

$$\Box^-_s = \frac{1}{\sqrt{2}}(\tau + \Delta)I_n + \frac{1}{\sqrt{2}}(2-n)\frac{\tau}{\Delta}I_n$$

$$+ \frac{1}{i\sqrt{2}\,\Delta}\left(\operatorname{Re} \sum_{j=1}^{n} \sum_{k=1}^{2n+1} \sigma(Z_j)_{\xi_k} \overline{\sigma(Z_j)}_{x_k}\right)I_n.$$

We need the following technical result.

8.13 Lemma. Let T_0 denote a classical pseudo-differential operator of order zero. Then

$$\sigma([\square^+, T_0]) = \varphi(x,0)(1 - \frac{\tau}{\Delta})\psi(\xi)\sigma(T_0)$$

$$+ \epsilon(\sigma(Z), \overline{\sigma(Z)})\,\sigma(T_{-1}) + \sigma(T_{-1}).$$

Proof. It suffices to consider a term of the form (typical of what occurs in the composition formula for symbols),

$$\frac{\partial}{\partial \xi_k}(\tau - \Delta)$$

$$= \tau_{\xi_k} - \frac{1}{\Delta}\Big(\tau\tau_{\xi_k} + \sum_{j=1}^{n}\big(\sigma(Z_j)_{\xi_k}\overline{\sigma(Z_j)} + \sigma(Z_j)\overline{\sigma(Z_j)}_{\xi_k}\big)\Big)$$

$$= \tau_{\xi_k}(1 - \frac{\tau}{\Delta}) + \epsilon(\sigma(Z), \overline{\sigma(Z)})\,\sigma(T_{-1}).$$

This proves Lemma 8.13.

8.14 Proposition. We have

(i) $\square^-\square^+$

$$= \varphi(x,0)\Big(\frac{1}{2}\sum_{j=1}^{n}(Z_j\overline{Z}_j + \overline{Z}_j Z_j) + (2-n)\,iT\Big)\varphi(x,0)I_n$$

$$+ \square^-([\overline{Z}_{n+1}(\varphi)]_0) + \epsilon(Z, \overline{Z})\,T_0 + T_0,$$

and

(ii) $\square^+\square^-$

$$= \varphi(x,0)\Big(\frac{1}{2}\sum_{j=1}^{n}(Z_j\overline{Z}_j + \overline{Z}_j Z_j) + (2-n)\,iT\Big)\varphi(x,0)I_n$$

$$+ [\overline{Z}_{n+1}(\varphi)]_0 \Box^- + \epsilon(Z, \overline{Z})T_0 + T_0.$$

Proof. We shall derive (i). The derivation of (ii) is similar. By Lemma 8.13 it suffices to work with the "raw symbols" $\Box_s^+$ and $\Box_s^-$. Then

$$\Box_s^- \circ \Box_s^+ = -\sum_{j=1}^{n} |\sigma(Z_j)|^2 I_n - (2-n)\tau I_n \tag{8.15}$$

$$+ \frac{1}{2i}\sum_{k=1}^{2n+1} (\tau+\Delta)_{\xi_k}(\tau-\Delta)_{x_k} I_n$$

$$- \frac{1}{i}\left(\mathrm{Re}\sum_{j=1}^{n}\sum_{k=1}^{2n+1} \sigma(Z_j)_{\xi_k}\overline{\sigma(Z_j)}_{x_k}\right)I_n$$

$$+ \left(1 - \frac{\tau}{\Delta}\right)\cdot\frac{1}{\sqrt{2}}(\tau+\Delta)\sigma(T_0) + \epsilon(\sigma(Z), \overline{\sigma(Z)})\sigma(T_0) + \sigma(T_0)$$

$$= \sigma\left(\left(\frac{1}{2}\sum_{j=1}^{n}(Z_j\overline{Z}_j + \overline{Z}_jZ_j) + (2-n)\,iT\right)I_n\right)$$

$$+ \frac{1}{2i}\left(\sum_{k=1}^{2n+1}(\tau+\Delta)_{\xi_k}(\tau-\Delta)_{x_k}\right)I_n$$

$$+ \epsilon(\sigma(Z), \overline{\sigma(Z)})\,\sigma(T_0) + \sigma(T_0).$$

Thus we are left with computing the sum $\sum_{k=1}^{2n+1} \ldots$. This yields

$$\frac{1}{2i}\sum_{k=1}^{2n+1}(\tau+\Delta)_{\xi_k}(\tau-\Delta)_{x_k}$$

$$= \frac{1}{2i}\sum_{k=1}^{2n+1}\tau_{\xi_k}\tau_{x_k} - \frac{1}{2i}\sum_{k=1}^{2n+1}\tau_{\xi_k}\Delta_{x_k}$$

$$+ \frac{1}{2i}\sum_{k=1}^{2n+1}\Delta_{\xi_k}\tau_{x_k} - \frac{1}{2i}\sum_{k=1}^{2n+1}\Delta_{\xi_k}\Delta_{x_k}$$

Now

$$\sum_{k=1}^{2n+1} \Delta_{\xi_k} \Delta_{x_k}$$

$$= \frac{1}{4} \Delta^{-2} \sum_{k=1}^{2n+1} (\Delta^2)_{\xi_k} (\Delta^2)_{x_k}$$

$$= \frac{1}{\Delta^2} \sum_{k=1}^{2n+1} \Big(\tau \tau_{\xi_k} + \sum_{j=1}^{n} (\sigma(Z_j)_{\xi_k} \overline{\sigma(Z_j)} + \sigma(Z_j) \overline{\sigma(Z_j)}_{\xi_k}) \Big)$$

$$\cdot \Big(\tau \tau_{x_k} + \sum_{j=1}^{n} (\sigma(Z_j)_{x_k} \overline{\sigma(Z_j)} + \sigma(Z_j) \overline{\sigma(Z_j)}_{x_k}) \Big).$$

Therefore

$$\frac{1}{2i} \sum_{k=1}^{2n+1} \tau_{\xi_k} \tau_{x_k} - \frac{1}{2i} \sum_{k=1}^{2n+1} \Delta_{\xi_k} \Delta_{x_k}$$

$$= \frac{1}{2i} \left(1 - \frac{\tau^2}{\Delta^2} \right) \sum_{k=1}^{2n+1} \tau_{\xi_k} \tau_{x_k} + \epsilon(\sigma(Z), \overline{\sigma(Z)}) \sigma(T_0)$$

$$= \epsilon(\sigma(Z), \overline{\sigma(Z)})\, \sigma(T_0).$$

Finally we consider

$$- \frac{1}{2i} \sum_{k=1}^{2n+1} \tau_{\xi_k} \Delta_{x_k} + \frac{1}{2i} \sum_{k=1}^{2n+1} \Delta_{\xi_k} \tau_{x_k}$$

$$= - \frac{1}{2i} \sum_{k=1}^{2n+1} \tau_{\xi_k} \Delta^{-1}$$

$$\cdot \Big(\tau \tau_{x_k} + \sum_{j=1}^{n} (\sigma(Z_j)_{x_k} \overline{\sigma(Z_j)} + \sigma(Z_j) \overline{\sigma(Z_j)}_{x_k}) \Big)$$

$$+\frac{1}{2i}\frac{1}{\Delta}\left(\tau\tau_{\xi_k}+\sum_{j=1}^{n}(\sigma(Z_j)_{\xi_k}\overline{\sigma(Z_j)}+\sigma(Z_j)\overline{\sigma(Z_j)}_{\xi_k})\right)\tau_{x_k}$$

$$=\varepsilon(\sigma(Z),\overline{\sigma(Z)})\,\sigma(T_0).$$

This proves Proposition 8.14.

We shall state a result, similar to Theorem 8.8, which shows, that $\Box^-$ can be defined invariantly, the way $\Box^+$ is defined. However, we shall not prove this since we are not going to make use of the proposition

8.16 Proposition. Let P_{out} denote the Poisson operator P on $\mathbb{R}_-^{2n+2}$, i.e., on the outside of $\overline{\mathbb{R}_+^{2n+2}}$.

Then

$$[\nu\bar{\partial}(P_{out})]_0=\Box^-+R,$$

where the pseudo-differential operator R has the following symbol

$$\sigma(R)=\varphi(x,0)\,\varepsilon\,(1+\frac{\tau}{\Delta})\,\psi(\xi)\,\sigma(T_0)$$

$$+\,\varepsilon(\sigma(Z),\overline{\sigma(Z)})\,\sigma(T_{-1})+\sigma(T_{-1}).$$

Again, φ and ψ are defined in Proposition 7.57 (ii), (iii).

The following is obvious from what we have already proved; the formula is stated for future reference.

8.17 Lemma.

$$\overline{Z}_{n+1}P=P\Box^++Q_0,$$

where Q_0 is an operator of Poisson type of order zero.

We recall (see (13.4) of [9]) that the Laplacian $\Box_b=\vartheta_b\bar{\partial}_b+\bar{\partial}_b\vartheta_b$

associated with the boundary Cauchy-Riemann complex has the following form (we only need its restriction to (0,1)-forms).

(8.18) $$\Box_b g = \sum_{j=1}^{n} \Big[\Big(-\frac{1}{2} \sum_{k=1}^{n} (Z_k \overline{Z}_k + \overline{Z}_k Z_k) + (n-2)iT \Big) g_j \Big] \overline{\omega}_j + \varepsilon(Zg, \overline{Z}g, g),$$

$g = \sum_{j=1}^{n} g_j \overline{\omega}_j$. Therefore we can rewrite Proposition 8.14 in the following form.

8.19 Proposition.

(i) $$\Box^- \Box^+ = -\varphi(x,0) \circ \Box_b \circ \varphi(x,0) + \Box^- \circ ([\overline{Z}_{n+1}(\varphi)]_0) + \varepsilon(Z, \overline{Z}) T_0 + T_0,$$

(ii) $$\Box^+ \Box^- = -\varphi(x,0) \circ \Box_b \circ \varphi(x,0) + [\overline{Z}_{n+1}(\varphi)]_0 \Box^- + \varepsilon(Z, \overline{Z}) T_0 + T_0$$

Chapter 9. A parametrix for $\square$ near bM; $n > 1$

In this chapter we shall construct an approximate local left inverse for the $\bar{\partial}$-Neumann problem when $n > 1$, i.e., we shall obtain a local representation of $u \in C^{\infty}_{(0,1)}(\overline{M})$ in terms of f, if

(9.1) $$\square(u) = f \text{ in } M,$$

(9.2) $$[\nu(u)]_{bM} = [\nu\bar{\partial}(u)]_{bM} = 0.$$

We shall next describe heuristically this local inverse (or approximate "Neumann operator"). In the brief description that follows we shall disregard error terms, (which will turn out to be smoothing operators), and pay no attention to the host of cut-off functions that must be used (which introduce additional error terms of smoothing operators). Thus according to Theorem 7.66 we have approximately

(9.3) $$u = G(f) + P([u]_0)$$

Now the first boundary condition in (9.2) gives the direct control of part of $[u]_0$ (namely $\nu(u)$). We will control the other part, u_t, indirectly in terms of f. To do this apply the boundary operator $B_{\bar{\partial}}$ (of the second boundary condition) to (9.3). Since $B_{\bar{\partial}} P[u_0] = \square^+ u_t$, we have approximately

(9.4) $$\square^+ u_t = -B_{\bar{\partial}} G(f)$$

Moreover $\square^- \square^+ = -\square_b$ approximately, where $\square_b$ is boundary analogue of the Laplacian on 1-forms. (See Proposition (8.19).) However when $n > 1$, $\square_b$ has an inverse K given by an integral operator, known quite explicitly; thus $K\Box_b = -I$ approximately. (It is here that the limitation $n > 1$ is required in this chapter.) Putting these things together in (9.4) gives that approximately

$$u_t = K\Box^{-} B_{\bar{\partial}} G(f), \quad \text{and hence}$$

to approximate inverse to (9.1) is then

$$(9.5) \qquad u = G(f) + P(K\Box^{-} B_{\bar{\partial}} G(f))$$

We now pass to the precise version of (9.5)

We use the notation of Chapters 7 and 8. Thus $U \supset\supset V$ are boundary neighborhoods of $\overline{M}$ in M', identified with subsets of $\mathbb{R}^{2n+2}$, such that, if $\mathbb{R}^{2n+2}$ has coordinates (x, ρ), $x = (x_1, \ldots, x_{2n+1})$, then $U_0 = U \cap bM$ is identified with a subset of $\mathbb{R}^{2n+1} = \{(x,\rho) \in \mathbb{R}^{2n+2};\ \rho = 0\}$. $\varphi(x, \rho)$ and $\psi(\xi)$ are the cut-off functions of Proposition 7.57(ii),(iii). According to Theorem (7.67), we can write

$$(9.3') \qquad \varphi^{(0)}u = G_1\,(\Box\ (\varphi^{(0)}\ u)) + P_1([\varphi^{(0)}u]_0),$$

if $\varphi^{(0)} \in C_0^\infty(V_1)$.

We shall now replace G_1 and P_1 by the "exact" Green's and Poisson operators G and P. This introduces smoothing operators (of infinite order) in the above. For simplicity of notation we shall not keep track of these resulting errors explicitly, but gather them together at the end of the proof of Lemma (9.11).

We shall compute $[\varphi^{(0)}u]_0$ in terms of f. We note, that by (7.61)

$$[\nu G\,(\Box\ (\varphi^{(0)}u))]_0 = 0$$

in $V_0 = V \cap \{(x,\rho) \in \mathbb{R}^{2n+2};\ \rho = 0\}$, and, since $[\nu(u)]_0 = 0$,

$$[\nu P([\varphi^{(0)} u]_0)]_0 = 0 \text{ in } V_0.$$

Next we recall some results of [9] concerning a parametrix for $\Box_b$, $n > 1$. Let $u(x,y) = (u_0(x,y), u_1(x,y), \ldots, u_{2n}(x,y))$ denote admissible coordinates in $U_0 = U \cap \mathbb{R}^{2n+1}$ (we refer to Chapter 4 for this notion). Define the kernel $\Phi_{n-2}(x,y)$ by

$$(9.6) \quad \Phi_{n-2}(x,y)$$

$$= \chi(x,y) \frac{(n-2)!}{2^{2-2n}\pi^{n+1}} (|u'(y,x)|^2 - iu_0(y,x))^{-n+1}$$

$$\cdot (|u'(y,x)|^2 + iu_0(y,x))^{-1} \cdot 2^{-n} |\det A(y)|^{-1},$$

where $\chi \in C_0^\infty(U_0 \times U_0)$, $\chi(y,x) = \chi(x,y)$ and $\chi(x,y) = 1$ on a neighborhood of the diagonal of $\overline{V}_0 \times \overline{V}_0$. We also set $u' = (u_1, \ldots, u_n)$ and the volume element $2^{-n}|\det A(y)|^{-1} dy$ is explained in Proposition 4.37.

We define the operator $K: C^\infty_{(0,1),0}(U_0) \longrightarrow C^\infty_{(0,1)}(U_0)$ by

$$(9.7) \quad Kh(x) = \sum_{j=1}^{n} \left(\int \Phi_{n-2}(x,y) h_j(y)\, dy\right) \overline{\omega}_j, \quad h \in C^\infty_{(0,1),0}(U_0).$$

<u>9.8 Proposition.</u> <u>If</u> $h \in C^\infty_{(0,1),0}(V_0)$, <u>then</u>

$$(9.9) \quad \Box_b K(h) = h + S_1(h)$$

$$(9.10) \quad K \circ \Box_b(h) = h + S_1'(h),$$

<u>where</u> S_1 <u>(or</u> S_1') <u>denotes an operator of type</u> 1. This is just Proposition 16.5 of [9], and shows that K is the approximate inverse for $\Box_b$.

Here, and in what follows S_m will systematically denote an operator of type m, in the sense of Folland-Stein [9], §15, 16. We shall return to these operators in Chapter 14. For the present we shall recall that operators of type m are smoothing of order m in the "good" directions. Thus the parametricies constructed in [9] for $\Box_b$ and $\bar{\partial}_b$ were operators of type 2 and 1, respectively. Also a vector field in the "good" directions composed with an S_m gives an S_{m-1}.

9.11 Lemma. Suppose $n > 1$. Let $V = V_1$ be the boundary neighborhood of $\overline{M}$ in M' defined in Theorem 7.62. Let $u \in C^{\infty}_{(0,1)}(\overline{\mathbb{R}^{2n+2}_+} \cap V)$, such that

$$\Box(u) = f \text{ in } V \cap \mathbb{R}^{2n+2}_+,$$

$$\nu(u) = \nu\bar{\partial}(u) = 0 \text{ in } V_0 = V \cap \{\rho = 0\}.$$

Let $\varphi^{(1)} \in C^{\infty}_0(V)$. Then there exist $\varphi^{(0)}, \varphi^{(-1)} \in C^{\infty}_0(V)$, such that

$$(9.12) \qquad \varphi^{(1)}u + \tilde{T}_{-1}(\varphi^{(0)}u)$$

$$+ P(S_1([(\tilde{T}_0(\varphi^{(-1)}u))_t]_0))$$

$$= G(\varphi^{(1)}f)$$

$$+ P(K \circ \Box^{-}[\varphi^{(1)}]_0\, B_{\bar{\partial}}(G(\varphi^{(0)}f))).$$

(Recall that $\sim$ indicates an extension via (7.61).)

Proof. We apply $B_{\bar{\partial}}$ to (9.3').

$$(9.13) \qquad B_{\bar{\partial}}(\varphi^{(0)}u) = [(\overline{Z}_{n+1}\varphi)u]_0$$

$$= B_{\bar{\partial}}(G(\Box(\varphi^{(0)}u))) + B_{\bar{\partial}}(P([\varphi^{(0)}u]_0)),$$

where we chose $\varphi^{(0)} \in C_0^\infty(V)$ so that $\varphi^{(1)} \subset\subset \varphi^{(0)}$.

By Definition 8.6 and Theorem 8.8

$$(9.14) \qquad B_{\bar\partial}(P([\varphi^{(0)}u]_0)) = B_{\bar\partial}(P([\varphi^{(0)}u_t]_0))$$
$$= \Box^+([\varphi^{(0)}u_t]_0).$$

Multiplying (9.13) by $\varphi^{(1)}$ it becomes

$$(9.15) \qquad 0 = [\varphi^{(1)}]_0 B_{\bar\partial}(G(\Box(\varphi^{(0)}u)))$$
$$+ [\varphi^{(1)}]_0 \Box^+([\varphi^{(0)}u_t]_0).$$

According to Lemma 8.13 if we interchange $\Box^+$ and $[\varphi^{(1)}]_0$, the error committed is negligible. Hence (9.15) becomes

$$(9.16) \qquad \Box^+ \circ [\varphi^{(1)}u_t]_0 + R([\varphi^{(0)}u_t]_0)$$
$$+ [\varphi^{(1)}]_0 B_{\bar\partial}(G(\Box(\varphi^{(0)}u))) = 0,$$

where R is a pseudo-differential operator induced by a symbol of the form

$$\varphi(x,0)(1 - \frac{\tau}{\Delta})\psi(\xi)\sigma(T_0)$$
$$+ \epsilon(\sigma(Z), \overline{\sigma(Z)})\,\sigma(T_{-1}) + \sigma(T_{-1}),$$

where φ and ψ were defined in Proposition 7.57 (ii), (iii).

We apply $\Box^-$ to (9.16) and use Proposition 8.19 (i) to obtain

$$(9.17) \qquad (-[\varphi]_0 \Box_b [\varphi]_0 + \epsilon(Z, \bar{Z})T_0 + T_0)([\varphi^{(1)}u_t]_0)$$
$$+ (\epsilon(Z, \bar{Z})T_0 + T_0)([\varphi^{(0)}u_t]_0)$$
$$+ \Box^- [\varphi^{(1)}]_0 B_{\bar\partial}(G(\Box(\varphi^{(0)}u))) = 0.$$

We apply K on the left. According to (9.10) we obtain

$$- (I + S_1)([\varphi^{(1)} u_t]_0) + S_1 T_0([\varphi^{(0)} u_t]_0)$$

$$+ K \circ \square^- [\varphi^{(1)}]_0 B_{\bar{\partial}} (G(\square(\varphi^{(0)} u))) = 0,$$

or

(9.18) $$[\varphi^{(1)} u_t]_0$$

$$= K \circ \square^- [\varphi^{(1)}]_0 B_{\bar{\partial}} (G(\square(\varphi^{(0)} u)))$$

$$+ S_1 \circ T_0([\varphi^{(0)} u_t]_0).$$

According to (9.3′)

(9.19) $$\varphi^{(1)} u = G(\square(\varphi^{(1)} u)) + P([\varphi^{(1)} u]_0).$$

We note that $[\varphi^{(1)} u]_0 = [\varphi^{(1)} u_t]_0 \oplus 0$. Therefore, substituting (9.18) into (9.19) we obtain

(9.20) $$\varphi^{(1)} u = G(\square(\varphi^{(1)} u))$$

$$+ P(K \circ \square^- [\varphi^{(1)}]_0 B_{\bar{\partial}} (G(\square(\varphi^{(0)} u))))$$

$$+ P(S_1 \circ T_0([\varphi^{(0)} u_t]_0)).$$

To obtain the formula given in the statement of the lemma we only need to simplify (9.20). Namely

$$\square(\varphi^{(1)} u) = \varphi^{(1)} f + T_1(\varphi^{(0)} u),$$

therefore

$$G(\square(\varphi^{(1)} u)) = G(\varphi^{(1)} f) + \tilde{T}_{-1}(\varphi^{(0)} u).$$

Similarly

$$[\varphi^{(1)}]_0 B_{\bar\partial} (G(\Box(\varphi^{(0)}u)))$$

$$= [\varphi^{(1)}]_0 B_{\bar\partial} (G(\varphi^{(0)}f))$$

$$+ [\varphi^{(1)}]_0[(T_{-1}(\varphi^{(-1)}u))_t]_0 + [(T_{-\infty}(\varphi^{(-1)}u))_t]_0,$$

where we set $\varphi \supset\supset \varphi^{(-1)} \supset\supset \varphi^{(0)} \supset\supset \varphi^{(1)}$. Collecting these formulas we obtain

(9.21) $$\varphi^{(1)}u + \tilde T_{-1}(\varphi^{(0)}u) + P(S_1 \circ T_0([\varphi^{(0)}u_t]_0))$$

$$+ P(K([(\tilde T_0(\varphi^{(-1)}u))_t]_0))$$

$$= G(\varphi^{(1)}f)$$

$$+ P(K \circ \Box^- [\varphi^{(1)}]_0 B_{\bar\partial} (G(\varphi^{(0)}f))).$$

Since K is certainly an S_1 we have derived Lemma 9.11.

We note that the cut-off functions are irrelevant from the point of view of the estimates. We make this more precise.

<u>9.22 Definition.</u> Let $n > 1$. Let V be as in Lemma 9.11. We say that the operator $N_a : C^\infty_{(0,1)}(V \cap \overline{\mathbb{R}^{2n+2}_+}) \longrightarrow C^\infty_{(0,1)}(V \cap \overline{\mathbb{R}^{2n+2}_+})$ is of <u>Neumann type</u> if it has the following form

(9.23) $$N_a(f) = G(\varphi_1 f)$$

$$+ P(K \circ \Box^- [\varphi_2]_0 B_{\bar\partial} (G(\varphi_3 f)))$$

for some φ_1, φ_2 and $\varphi_3 \in C^\infty_0(V)$.

<u>9.24 Definition.</u> Let V be the boundary neighborhood given in Lemma

9.11. We say that the operator $R: C^{\infty}_{(0,1)}(\overline{V \cap \mathbb{R}^{2n+2}_{+}}) \longrightarrow C^{\infty}_{(0,1)}(\overline{V \cap \mathbb{R}^{2n+2}_{+}})$ is of remainder type if it can be written in the following form

$$\text{(9.25)} \qquad R(u) = \tilde{T}_{-1}(\varphi_1 u) + \varphi_2 P(S_1 T_0[(\tilde{T}_0(\varphi_3 u))_t]_0)$$

with $\varphi_1, \varphi_2, \varphi_3 \in C^{\infty}_0(V)$.

We restate Lemma 9.11 as follows.

9.26 Proposition. Suppose $n > 1$. Let V be as in Theorem 7.62. Let $\varphi^{(1)} \in C^{\infty}_0(V)$. Then there exist operators N_a of Neumann type and R of remainder type, such that

$$\varphi^{(1)} u - R(u) = N_a(f),$$

whenever $u \in C^{\infty}_{(0,1)}(\overline{V \cap \mathbb{R}^{2n+2}_{+}})$ and

$$\Box(u) = f \text{ in } V \cap \mathbb{R}^{2n+2}_{+},$$

$$\nu(u) = \nu\bar{\partial}(u) = 0 \text{ in } V_0.$$

Notational Remark

It might be well to record once more the definition we have used in this chapter, and which we will continue to use in further chapters, namely that of operators S_m and T_k. We denote by S_m the (Heisenberg-group-type) operators of type m; T_k is a standard pseudo-differential operator of order k.

Chapter 10. The parametrix for $\square$ near bM; n=1

In this case

$$\square_b(\varphi) = -\frac{1}{2}(Z_1\overline{Z}_1 + \overline{Z}_1 Z_1)(\varphi) - iT(\varphi) \tag{10.1}$$

$$+ \epsilon(Z_1, \overline{Z}_1, \varphi)$$

$$= -\overline{Z}_1 Z_1(\varphi) + \epsilon(Z_1, \overline{Z}_1, \varphi)$$

Here we have the added complication that $\square_b$ has no inverse; i.e., parametrix. However, we can make use of the result concerning the solvability of the Lewy equation discussed in Chapter 3. The idea, stated somewhat imprecisely, is as follows:

We let $\overline{C}_b$ denote the (presumptive) projection on the boundary values of anti-holomorphic functions. Then on the orthogonal complement, (whose "projection" is given by $I - \overline{C}_b$), $\square_b$ has an inverse; namely by the results of Chapter 3 we can find an integral operator $\overline{K}$ so that $\overline{K}\square_b = I - \overline{C}_b$, approximately; and so $\square^+$ has an inverse on the image of $I - \overline{C}_b$, given by $-\overline{K}\square^-$. On the other hand let E^+ be the "projection" operator (given by an ordinary pseudo-differential operator of order 0) corresponding to a conic neighborhood of the characteristic variety of $\square^+$. The important fact is that $E^+\overline{C}_b = \overline{C}_b E^+ = 0$, approximately, and hence the projection $\overline{C}_b$ is subordinate to projection $E^- = I - E^+$; moreover $\square^+$ is elliptic away from its characteristic variety. Thus there exists a pseudo-differential operator of order -1, Q_{E^-}, so that $Q_{E^-}\square^+ = E^-$ approximately. It is then easy to see that $-E^+\overline{K}\square^- + Q_{E^-}$ is an approximate inverse to $\Box^{+}$.

An alternative form for this is

$$-\overline{K}\Box^{-} + \overline{C}_b Q_{E^-}$$

We set

(10.2) $$\overline{S}_\epsilon(z,t) = \frac{2}{\pi^2}(|z|^2 + \epsilon^2 + it)^{-2}, \quad z = x + iy,$$

and we define

$$\overline{C}_b(f) = \lim_{\epsilon \to 0} f * \overline{S}_\epsilon,$$

where the convolution is with respect to the Heisenberg group H_1. Let $\overline{K}$ denote the operator given by convolution on H_1 with the function

(10.3) $$\overline{\Phi}(z,t) = \pi^{-2} \log\left(\frac{|z|^2 + it}{|z|^2 - it}\right)(|z|^2 + it)^{-1}.$$

<u>(10.4) Lemma</u>. <u>We set</u> $L = \frac{\partial}{\partial z} + i\overline{z}\frac{\partial}{\partial t}$. <u>Then</u>

(10.5) $$(-\overline{L}L)\overline{K} = \overline{K}(-\overline{L}L) = I - \overline{C}_b.$$

<u>Proof</u>. This is a consequence of Lemma 3.18, on taking complex conjugates.

We return to $\overline{M} \subset M'$. Let U be a boundary neighborhood of $\overline{M}$ in M' and let $u = u(x,y)$ denote admissible coordinates in $U_0 = bM \cap U$ as given in Proposition 4.3.

Let $V \subset\subset U$ as in Proposition 9.8. Set

(10.6) $$\overline{C}_{b,u}(f)(x) = \lim_{\epsilon \to 0} \int_{U_0} \chi(y,x)\overline{S}_\epsilon(u(y,x))f(y)\,d(bM)(y),$$

where $\chi \in C_0^\infty(U \times U)$, $\equiv 1$ on a neighborhood of the diagonal of $V_0 \times V_0$, and $\chi(y,x) = \chi(x,y)$, and

(10.7) $$d(bM)(y) = \frac{1}{2}|\det A(y)|^{-1}\, dy.$$

Recall that if $Z_1 = \frac{1}{2}(V_1 - iV_2)$ and $V = (V_1, V_2)$, then

(10.8) $$V = A(y)\frac{\partial}{\partial y}$$

(see Proposition 4.3). We set

$$\overline{K}_u(f)(x) = \int_{U_0} \chi(y,x)\overline{\Phi}(u'(y,x), u_0(y,x))f(y)\, d\,(bM)(y),$$

where we set $u' = (u_1, u_2) = u_1 + iu_2$.

<u>(10.9) Proposition.</u> <u>Let</u> V <u>be the boundary neighborhood given in Proposition 9.8. Then</u>

(10.10) $$\overline{K}_u \circ \Box_b = I - \overline{C}_{b,u} + S_1,$$

(10.11) $$\Box_b \circ \overline{K}_u = I - \overline{C}_{b,u} + S_1$$

on V_0.

<u>Proof.</u> The result follows from Lemma 10.4 modulo the usual changes necessary to transfer arguments from the Heisenberg group to a strongly pseudo-convex CR manifold (e.g. see the proof of Proposition 16.2 of [9]).

From now on, to simplify the notation we shall drop the u subscripts from $\overline{K}_u$ and $\overline{C}_{b,u}$.

Let $\langle u(y,x), \sigma\rangle = \langle x-y, \xi\rangle$. This defines σ as a linear function of

ξ with parameters x and y, i.e., $\sigma = \sigma(x, \xi, y)$.

10.12 Lemma. Let $\psi(\sigma) \in C^{\infty}(\mathbb{R}^3)$, $\equiv 0$ in $|\sigma| < 1$ and $\equiv 1$ if $|\sigma| > 2$. We set

(10.13)
$$\chi^{(-)}(\sigma) = \begin{cases} 0, \sigma > 0 \\ 1, \sigma \leqq 0. \end{cases}$$

Then $\overline{C}_b = \overline{C}_{b,1} + T_{-\infty}$, where

(10.14)
$$\sigma(\overline{C}_{b,1}) = 2(\chi^{(-)}\psi)(\sigma(x,\xi,y)) \exp\left(\frac{\sum_{j=1}^{2} \sigma_j(x,\xi,y)^2}{4\sigma_0(x,\xi,y)}\right) \cdot (1 + 0(|x-y|)).$$

Proof. Conjugating Proposition 3.27 we get

(10.15)
$$\frac{2}{\pi^2}(\epsilon^2 + |u'(y,x)|^2 + iu_0(y,x))^{-2} = (2\pi)^{-3}\int_{\mathbb{R}^3} 4\chi^{(-)}(\sigma) e^{\epsilon^2\sigma_0 + \frac{\sigma_1^2+\sigma_2^2}{4\sigma_0} + i\langle u(y,x), \sigma\rangle} d\sigma$$

We note that (a) the integral in (10.15) converges absolutely, and (b)

$$\chi^{(-)}(\sigma)\exp\left(\epsilon^2\sigma_0 + \frac{\sigma_1^2+\sigma_2^2}{4\sigma_0}\right) \in C^{\infty}(\mathbb{R}^3 - \{0\}).$$

Let $\overline{C}_{b,1}(f) = \lim_{\epsilon \to 0} \overline{C}_{b,1,\epsilon}(f)$, where

$$\overline{C}_{b,1,\epsilon}(f) = \int_{U_0} \chi(x,y)\overline{C}_{b,1,\epsilon}(x,y) f(y)\, d(bM)(y),$$

where

(10.16) $\overline{C}_{b,1,\varepsilon}(x,y)$

$$= (2\pi)^{-3} \int_{\mathbb{R}^3} 4(\chi^{(-)}\psi)(\sigma) e^{\varepsilon^2 \sigma_0 + \frac{\sigma_1^2 + \sigma_2^2}{4\sigma_0} + i\langle u(y,x), \sigma\rangle} d\sigma$$

It is easy to see that $\overline{C}_b - \overline{C}_{b,1}$ is an integral operator with a C^∞ kernel therefore it represents a $T_{-\infty}$ operator. To obtain $\sigma(\overline{C}_{b,1})$ we shall compute $\sigma(\overline{C}_{b,1,\varepsilon})$. We begin by changing the variables of integration in (10.15). According to Proposition 4.3

$$u_j(y,x) = \sum_{k=0}^{2} B_{jk}(y)(x_k - y_k) + O(|x-y|^2),$$

where ${}^tB(y) = A(y)^{-1}$. Therefore

$$\begin{aligned}\langle u(y,x), \sigma\rangle &= \langle (B(y) + O(|x-y|))(x-y), \sigma\rangle \\ &= \langle x-y, ({}^tB(y) + O(|x-y|))\sigma\rangle \\ &= \langle x-y, \xi\rangle,\end{aligned}$$

where we set

(10.17) $$\xi = ({}^tB(y) + O(|x-y|))\sigma .$$

Therefore

(10.18) $$\sigma(x,\xi,y) = (A(y) + O(|x-y|))\xi ,$$

and the change of variables $\sigma \longrightarrow \xi$ yields

$$\overline{C}_{b,1,\varepsilon}(x,y)$$
$$= (2\pi)^{-3} \int_{\mathbb{R}^3} 4(\chi^{(-)}\psi)(\sigma(x,\xi,y)) e^{i\langle x-y, \xi\rangle}$$

$$\cdot \exp\left(\varepsilon^2 \sigma_0(x,\xi,y) + \frac{\sum_{j=1}^{2} \sigma_j(x,\xi,y)^2}{4\sigma_0(x,\xi,y)}\right)\left|\frac{\partial \sigma(x,\xi,y)}{\partial \xi}\right| d\xi,$$

and since $d(bM)(y) = \frac{1}{2}|\det A(y)|^{-1}\, dy$, letting $\varepsilon \longrightarrow 0$ we obtain (10.14). This yields Lemma 10.12.

<u>10.19 Lemma.</u> $\sigma(\overline{C}_{b,1}) \in S^0_{1/2,1/2}$.

<u>Proof.</u> Let $\sigma' = (\sigma_1, \sigma_2)$. A simple computation yields

$$\left|\partial^\alpha_{\sigma'} \partial^j_{\sigma_0} e^{|\sigma'|^2/4\sigma_0}\right|$$

$$\leqq |\sigma_0|^{-|\alpha|/2-j} \sum_i a_i \left(\frac{|\sigma'|^2}{4\sigma_0}\right)^{p_i} e^{|\sigma'|^2/4\sigma_0},$$

if $\sigma_0 < 0$. The p_i's are nonnegative integers and depend only on $|\alpha|$ and j.

Since

$$|x^k e^x| \leqq k^k e^{-k} \quad \text{if } x < 0,$$

we have

$$|\sigma_0|^{|\alpha|/2+j} \left|\partial^\alpha_{\sigma'} \partial^j_{\sigma_0} e^{|\sigma'|^2/4\sigma_0}\right| \leqq C_{|\alpha|,j}.$$

Similarly

$$|\sigma'|^{|\alpha|+2j} \left|\partial^\alpha_{\sigma'} \partial^j_{\sigma_0} e^{|\sigma'|^2/4\sigma_0}\right| \leqq C_{|\alpha|,j}.$$

Therefore

(10.20)
$$\left|\partial_{\sigma'}^{\alpha}\partial_{\sigma_0}^{j} e^{|\sigma'|^2/4\sigma_0}\right| \leqq C_{|\alpha|,j}\left(|\sigma'|^{-(|\alpha|+2j)} + |\sigma_0|^{-(|\alpha|/2+j)}\right).$$

Since $\psi = 0$ near the origin (10.20) implies

$$\left|\partial_{\sigma'}^{\alpha}\partial_{\sigma_0}^{j}\left((\psi\chi^{(-)})(\sigma)\, e^{|\sigma'|^2/4\sigma_0}\right)\right| \leqq C_{|\alpha|,j}(1+|\sigma'|^2+|\sigma_0|)^{-|\alpha|/2-j}.$$

If $|\sigma'|^2 + |\sigma_0| > 1$, then

$$2(1+|\sigma'|^2+|\sigma_0|) > 1+|\sigma'|+|\sigma_0|.$$

Therefore we obtain the following, rather crude, estimate, which, nevertheless suffices for our purposes. Namely

(10.21)
$$\left|\partial_{\sigma'}^{\alpha}\partial_{\sigma_0}^{j}\left((\chi^{(-)}\psi)(\sigma)\, e^{|\sigma'|^2/4\sigma_0}\right)\right| \leqq C_{|\alpha|,j}(1+|\sigma|)^{-|\alpha|/2-j} \leqq C_{|\alpha|,j}(1+|\sigma|)^{-|\alpha|/2-j/2}.$$

Now we can easily prove Lemma 10.19. If $\sigma = \sigma(x,\xi,y)$, then

$$\frac{\partial}{\partial\xi} = O(1)\frac{\partial}{\partial\sigma} \quad \text{as } |\xi| \longrightarrow \infty,$$

$$\frac{\partial}{\partial x} = O(|\xi|)\frac{\partial}{\partial\sigma} \quad \text{as } |\xi| \longrightarrow \infty,$$

$$\frac{\partial}{\partial y} = O(|\xi|)\frac{\partial}{\partial\sigma} \quad \text{as } |\xi| \longrightarrow \infty.$$

Therefore

$$\left|\partial_\xi^\alpha \partial_x^\beta \partial_y^\gamma \left((\chi^{(-)}\psi)(\sigma(x,\xi,y)\right) \exp\left(\frac{\sum_{j=1}^{2} \sigma_j(x,\xi,y)^2}{4\sigma_0(x,\xi,y)}\right)\right)\right|$$

$$\leq C_{\alpha,\beta,\gamma}(1+|\xi|)^{-|\alpha|/2+|\beta|/2+|\gamma|/2},$$

which proves Lemma 10.19.

Next we construct the inverse for $\square^+$. Recall that the principal symbol of $\square^+$ is $\frac{1}{\sqrt{2}}(\tau - \Delta)$, where $\Delta = (\tau^2 + 2|\sigma(Z_1)|^2)^{1/2}$. Therefore, the characteristic variety $\Sigma^{(+)}$ of $\square^+$ is the set

$$\{\tau > 0;\ \sigma(Z_1)(x,\xi) = 0\}$$

$$= \left\{\sum_{k=0}^{2} A_{0k}(x)\,\xi_k > 0 \text{ and } \sum_{k=0}^{2} A_{jk}(x)\xi_k = 0,\quad j=1,2\right\}.$$

The idea of the construction is that in some conic neighborhood of $\Sigma^{(+)}$, $\square_b$ has an inverse by virtue of Proposition 10.9 and Lemma 10.12; on the other hand, $\square^+$ is elliptic, hence invertible, away from $\Sigma^{(+)}$. To be more precise

$$\operatorname{supp} \chi^{(-)}(\sigma(x,\xi,y))$$

$$\subset \{\xi \in \mathbb{R}^3;\ \sigma_0(x,\xi,y) < 0\},$$

where

$$\sigma_0(x,\xi,y) = \sum_{k=0}^{2} A_{0k}(y)\xi_k + O(|x-y|\,\xi$$

$$= \sum_{k=0}^{2} A_{0k}(0)\xi_k + O(|x| + |y|)\xi.$$

Let $x, y \in U_0$, U_0 a neighborhood of the origin in $\mathbb{R}^3$. Since $A(0)$ is nonsingular $\left\{ \sum_{k=0}^{2} A_{0k}(0)\xi_k < 0 \right\}$ defines a half-space in $\mathbb{R}^3$. Therefore, by choosing U_0 sufficiently small, the complement of

$$\bigcup_{x,y \in U_0} \{\xi \in \mathbb{R}^3 ; \sigma_0(x, \xi, y) < 0\}$$

in $\mathbb{R}^3$ contains a conic neighborhood, $A(0)^{-1} W^{(+)}$, where $W^{(+)}$ is a conic neighborhood of the ray $\{(\sigma_0, 0, 0); \sigma_0 > 0\}$. We note that $A(0)^{-1}\{(\sigma_0, 0, 0), \sigma_0 > 0\}$ is the characteristic variety of $\Box^+$ with $x = 0$. We choose U_0 so small that $\Box^+$ is elliptic outside of the conic set $A(0)^{-1} W_1^{(+)}$, for some $W_1^{(+)} \subset W^{(+)}$, when $x, y \in U_0$.

Let $\varphi^{(+)}(\sigma) = \varphi^{(+)}(\sigma_0, \sigma_1, \sigma_2)$ be C^∞ in $\mathbb{R}^3$, homogeneous of degree zero for large $|\xi|$, identically one on some neighborhood of $W_1^{(+)}$ and vanish in the complement of $W^{(+)}$. Set $\varphi^{(-)}(\sigma) = 1 - \varphi^{(+)}(\sigma)$. Let E^+ and E^- denote the pseudo-differential operators induced by the $S^0_{1,0}$ symbols $\varphi^{(+)}(A(0)\xi)$ and $\varphi^{(-)}(A(0)\xi)$, respectively. Then the complement of

$$\bigcup_{x,y \in U_0} \{\xi \in \mathbb{R}^3 ; \sigma_0(x, \xi, y) < 0\}$$

contains the support of $\varphi^{(+)}(A(0)\xi)$ in its interior. Furthermore, E^+, with symbol in $S^0_{1,0}$ and $\overline{C}_b$, with symbol in $S^0_{1/2,1/2}$ can be combined via the calculus of pseudo-differential operators (see Hörmander [ ]). Hence we have the following result.

<u>10.22 Lemma</u>. $\sigma(E^+ \circ \overline{C}_b)$ <u>and</u> $\sigma(\overline{C}_b \circ E^+)$ <u>both belong to</u> $S^{-\infty}_{1/2,1/2}$,

i.e., $E^+ \overline{C}_b$ and $\overline{C}_b \circ E^+$ are both integral operators with C^∞ kernel.

Now we define the left inverse, or parametrix, K^+, for $\Box^+$, as follows.

(10.23) $$K^+ = -Q_{E^-} + E^+ \overline{K} \circ \Box^-,$$

where Q_{E^-} denotes the parametrix for $\Box^+$ in the support of E^-, i.e.,

(10.24) $$Q_{E^-} \circ \Box^+ = E^- + T_{-\infty}.$$

10.25 Lemma.

$$K^+ \circ \Box^+ = -I + T_0 S_1 T_0 \text{ in } V_0.$$

Proof.

$$K^+ \circ \Box^+ = -Q_{E^-} \circ \Box^+ + E^+ \circ \overline{K} \circ \Box^- \circ \Box^+$$

$$= -E^- + T_{-\infty} + E^+ \overline{K}(-\Box_b + \epsilon(Z, \overline{Z}) T_0 + T_0)$$

$$= -E^- + T_{-\infty} - E^+ (I - \overline{C}_b + S_1 T_0)$$

$$= -I + T_0 S_1 T_0 + T_{-\infty},$$

where we used Lemma 10.22. Notice that $T_{-\infty}$ can always be put in the form of $S_1 T_0$. This proves Lemma 10.25.

10.26 Lemma. Suppose n=1. Let $V = V_1$ be the boundary neighborhood of $\overline{M}$ in M' defined in Theorem 7.62. Let $u \in C^\infty_{(0,1)}(\overline{\mathbb{R}^4_+} \cap V)$ such that

$$\Box(u) = f \text{ in } V \cap \mathbb{R}^4_+$$

$$\nu(u) = \nu\bar{\partial}(u) = 0 \text{ in } V_0 = V \cap \{\rho = 0\}.$$

Let $\varphi^{(1)} \in C^\infty_0(V)$. Then there exists $\varphi^{(0)}, \varphi^{(-1)} \in C^\infty_0(V)$, such that

(10.27)
$$\varphi^{(1)}u + \widetilde{T}_{-1}(\varphi^{(0)}u)$$
$$+ P((S_1 T_0)([\widetilde{T}_0(\varphi^{(-)}u)]_t [_0))$$
$$= G(\varphi^{(1)}f)$$
$$+ P(K^+[\varphi^{(1)}]_0 B_{\bar{\partial}}(G(\varphi^{(0)}f))).$$

Proof. The proof is similar to the proof of Lemma 9.11 and we will not repeat it.

10.28 Definition. Let n=1. Let V be as in Lemma 10.26. We say that the operator $N_a : C^\infty_{(0,1)}(V \cap \overline{\mathbb{R}^4_+}) \longrightarrow C^\infty_{(0,1)}(V \cap \overline{\mathbb{R}^4_+})$ is of Neumann type if it has the following form

(10.29)
$$N_a(f) = G(\varphi_1 f)$$
$$+ P(K^+[\varphi_2]_0 B_{\bar{\partial}}(G(\varphi_3 f))),$$

with $\varphi_1, \varphi_2, \varphi_3 \in C^\infty_0(V)$.

10.30 Proposition. Proposition 9.26 holds when $n = 1$.

We shall see that $\overline{C}_b Q_{E^-} - \overline{K}\square^-$ gives an alternative approximate inverse to $\square^+$, in the following sense (see also (10.25)).

10.31 Lemma

$$(\overline{C}_b Q_{E^-} - \overline{K}\square^-)\square^+ = I + S_1 T_0 \text{ in } V_0.$$

Proof. $\overline{K}\square^- \square^+ = -\overline{K}(\square_b + \text{error})$
$$= -I + \overline{C}_b + S_1 T_0$$

using (8.14) and (10.10). By (10.24 however, $\overline{C}_b Q_{E^-} \overline{K} = \overline{C}_b E^- + T_{-\infty}$, and the latter equals $\overline{C}_b + T_{-\infty}$, because of (10.22). Putting these together proves the lemma.

Remark. If we use Lemma (10.31), an approximate Neumann operator N_a can be written as

(10.32) $$G(f) + P(\overline{K}\Box^{-}[\varphi_2]_0 B_{\overline{\partial}} G(\varphi_3 f))$$

plus an error term involving $\overline{C}_b Q_{E^-}$.

Now the form (10.32) is the same as the corresponding form (9.23) for N_a, when $n > 1$. The error term will be

(10.33) $$-P(\overline{C}_b Q_{E^-}[\varphi_2]_0 B_{\overline{\partial}} G(\varphi_3 f))$$

which will be easier to handle than (10.32).

Part III. The Estimates

Guide to Part III

Here we deal with the regularity properties of the solutions of the $\bar{\partial}$-Neumann problem, (11.1) and (11.2) below.

L^2 theory

The L^2 theory of Kohn allows one to write down an "abstract" solution to the problem, in terms of a "Neumann operator" N. The theory proceeds by using L^2 Sobolev estimates, and culminates in the assertion that $Nf \in C^\infty(\overline{M})$ if $f \in C^\infty(\overline{M})$. These results are reviewed, with no proofs given, in Chapter 11. We then turn to estimates for N in other function spaces.

Several types of operators

Using the construction of approximate Neumann operators carried out in Chapters 9 and 10, the problem can be reduced to finding estimates for the following classes of operators:

(i) The restriction operator, mapping functions of M to functions on bM.

(ii) Pseudo-differential operators of the standard kind acting on functions on bM (or M). These are denoted generically by T.

(iii) Integral operators, related to the convolution operators on the Heisenberg group, acting on functions on bM. These are denoted generically by S.

(iv) Poisson operators, carrying functions on bM to functions on M.

Estimates for these operators

The required estimates are carried out, for the most part, in Chapters 12 - 14. Many of these estimates are either "classical" or previously known. Nevertheless we present many of the details here, enough to give a complete view of the subject. Besides estimates of the type arising for (iii), the novel results are certain estimates for Poisson operators. The former are described in Theorem 14.3, and the latter in Theorem 14.4 and Lemma 15.34.

Estimates for the $\bar{\partial}$-Neumann problem

Chapter 15 contains our main results, which are obtained as an application of the above estimates. Thus there is always a gain of one in all directions, a gain ot two in the "good" directions, and also a gain of two in the normal anti-holomorphic direction.

The equation $\bar{\partial} U = f$

The Neumann operator leads to a solution of $\bar{\partial} U = f$ (whenever this is possible), which solution is orthogonal to holomorphic functions. The estimates for this solution are essentially consequences of the corresponding results for the $\bar{\partial}$-Neumann problem, but there are certain additional obstacles that must be surmounted. The results are (roughly) a gain of one-half in all directions, and a gain of one in the "good" directions The details are in Chapter 16.

Chapter 11. Review of the L^2 theory

We shall now summarize the results of Kohn [21], (see also the exposition in [8]), concerning the L^2 and regularity theory for the solution of the $\bar{\partial}$-Neumann problem.

As before, M is an open sub-domain in a larger complex manifold M′; M has a smooth boundary, bM, which is strongly pseudo-convex. $C^\infty_{(0,1)}(\overline{M})$ denotes the (0,1) forms in $\overline{M}$ which are C^∞ up to the boundary; $L^2_{(0,1)}(\overline{M})$ denotes its closure in the L^2 norm (using the Levi-metric of Chapter 5). We are concerned, in effect, with the problem of solving

$$(11.1) \qquad \Box u = f$$

with the boundary conditions

$$(11.2) \qquad \nu(u)\big|_{bM} = 0, \quad \text{and} \quad \nu(\bar{\partial}u)\big|_{bM} = 0$$

where u and f are (0,1) forms.

The following theorem describes the L^2 and C^∞ theory.

11.3 Theorem. $\Box$, originally defined on those $u \in C^\infty_{(0,1)}(\overline{M})$ which satisfy the boundary conditions (11.2), has a unique extension to a self-adjoint (unbounded) operator $\Box_e$ on $L^2_{(0,1)}(\overline{M})$, satisfying the following:

(i) If $u \in \text{domain}(\Box_e) \cap C^\infty_{(0,1)}(\overline{M})$, then u satisfies the boundary conditions (11.2).

(ii) Let $\mathcal{H}$ = null space of $\Box_e$, (the "harmonic" space). Then $\mathcal{H}$ is finite-dimensional, and consists of elements that belong to $C^\infty_{(0,1)}(\overline{M})$. Moreover $L^2_{(0,1)}(\overline{M}) = \text{range}(\Box_e) \oplus \mathcal{H}$.

(iii) Suppose N is defined as $N(f) = u$, where $\Box_e(u) = f$, $u \perp \mathcal{H}$, and $N(f) = 0$, if $f \in \mathcal{H}$. Then $f \longrightarrow N(f)$ is a bounded operator on L^2.

(iv) $N(f) \in C^{\infty}_{(0,1)}(\overline{M})$, if $f \in C^{\infty}_{(0,1)}(\overline{M})$.

The operator N is called the Neumann operator.

Remarks. Observe that if $f \in L^2_{(0,1)}(\overline{M})$ and f is orthogonal to $\mathcal{H}$, then $u = N(f)$ gives a weak solution of our original problem ((11.1) and (11.2)) in the sense that

$$(11.4) \qquad (u, \Box\varphi) = (f, \varphi)$$

where φ is any element in $C^{\infty}_{(0,1)}(\overline{M})$ which satisfies the boundary conditions (11.2). Moreover u is the unique weak-solution, modulo elements of $\mathcal{H}$. If also $f \in C^{\infty}_{(0,1)}(\overline{M})$, then we get a solution to (11.1) and (11.2) in the usual sense.

Taking for granted Theorem 11.3, our main task then will be to prove the regularity results for the Neumann operator N, refining conclusions (iii) and (iv) to various function spaces.

The study of these function spaces is the subject of Chapters 12-14. We return to the Neumann operator proper in Chapter 15.

Note: In carrying out the estimates for $N(f)$ in Chapters 15 and 16 we shall assume that f is orthogonal to the harmonic space; the complement is a finite dimensional space of smooth forms, and hence this represents no limitation on the validity of our results.

Chapter 12. The Besov spaces $B^p(\mathbb{R}^m)$

We shall consider the Besov spaces, denoted by $\Lambda^{p,p}_{1-1/p}(\mathbb{R}^m)$ in [33], Chapter V, §5.* Here $1 < p < \infty$. We shall always write $B^p(\mathbb{R}^m) = \Lambda^{p,p}_{1-1/p}(\mathbb{R}^m)$. For simplicity of notation we have written $m = 2n+1$.

a. Definition

The space $B^p(\mathbb{R}^m)$ consists of all $f \in L^p$ for which the norm

$$\|f\|_p + \Big(\int_{\mathbb{R}^m} \frac{\|f(x-t) - f(x)\|_p^p}{|t|^{m+p-1}}\, dt\Big)^{1/p}$$

is finite. (Here $\|f(x-t) - f(x)\|_p^p = \int_{\mathbb{R}^m} |f(x-t) - f(x)|^p dx$.)

For our applications we shall need an equivalent characterization.

b. A characterization: Rate of approximation

Suppose $f \in B^p(\mathbb{R}^m)$; then there exists a family $\{f_\varepsilon\}_{0<\varepsilon\le 1}$ of smooth functions so that $f_\varepsilon \longrightarrow f$ in L^p norm as $\varepsilon \longrightarrow 0$, at a definite rate, while $\|\nabla f_\varepsilon\|_p = \|(\sum_{j=1}^m |\frac{\partial f_\varepsilon}{\partial x_j}|^2)^{1/2}\|_p$ can be controlled appropriately, as $\varepsilon \longrightarrow 0$. More precisely:

(12.1) Lemma. *Suppose* $f \in B^p$. *Then* $\exists\, f_\varepsilon \in C^\infty(\mathbb{R}^m)$, $0 < \varepsilon \le 1$ *so that*

---

* See the bibliographical remarks at the end of Chapter 13.

$$(12.2) \qquad \int_0^1 \|f-f_\epsilon\|_p^p \epsilon^{-p} d\epsilon < \infty \quad \text{and} \quad \int_0^1 \|\nabla f_\epsilon\|_p^p d\epsilon < \infty.$$

Conversely, if $\exists f_\epsilon$, so that (12.2) holds, then $f \in B^p$. The B^p norm of f is equivalent with the pth root of the sum of $\|f\|_p$ and the two quantities in (12.2).

Proof. Let $u(x,y)$ be the Poisson integral of f, (according to Chapter V, of [33]) then we know that

$$(12.3) \qquad \int_0^\infty \|\frac{\partial u}{\partial y}\|_p^p dy < \infty \quad \text{and} \quad \int_0^\infty \|\frac{\partial u}{\partial x}\|_p^p dy < \infty$$

by (61) and (62) of that chapter.

Let $f_\epsilon(x) = u(x,\epsilon)$. The second inequality of (12.3) proves the second inequality of (12.2). Moreover

$$u(x,0) - u(x,\epsilon) = f(x) - f_\epsilon(x) = -\int_0^\epsilon \frac{\partial u}{\partial y} dy, \quad \text{so}$$

$$\|f-f_\epsilon\|_p \leq \int_0^\epsilon \|\frac{\partial u}{\partial y}\|_p dy. \quad \text{However}$$

$$\int_0^1 (\epsilon^{-1} \int_0^\epsilon \varphi(y)dy)^p d\epsilon \leq (p/(p-1)) \int_0^\infty (\varphi(y))^p dy.$$

(This is Hardy's inequality), and so the first inequality of (12.2) follows from the first inequality of (12.3).

Conversely suppose that (12.2) is satisfied. Write

$$f = (f-f_\epsilon) + f_\epsilon$$

and

$$f(x-t) - f(x) = \{f(x-t) - f_\varepsilon(x-t)\} - \{f(x) - f_\varepsilon(x)\}$$

$$+ \{f_\varepsilon(x-t) - f_\varepsilon(x)\}$$

Now take $\varepsilon = |t|$,

and use the fact that $\|f_\varepsilon(x-t) - f_\varepsilon(x)\|_p \leq C|t| \, \|\nabla f_\varepsilon\|_p$.

Then $$\int_{|t|\leq 1} \|f(x-t) - f(x)\|_p^p \frac{dt}{|t|^{m+p-1}} \leq C \int_0^1 \|f-f_\varepsilon\|_p^p \varepsilon^{-p} d\varepsilon$$

$$+ C \int_0^1 \|\nabla f_\varepsilon\|_p^p d\varepsilon < \infty .$$

The finiteness of $\int_{|t|\geq 1} \|f(x-t) - f(x)\|_p^p \frac{dt}{|t|^{m+p-1}}$ is obvious from the fact that $f \in L^p$. The lemma is proved.

c. The space $L_1^p(\mathbb{R}_+^{m+1})$

We consider the half-space $\mathbb{R}_+^{m+1}$, with boundary $\mathbb{R}^m$. It will be useful to write $(x,y) = (x_1, \ldots, x_m, y)$ as coordinates for points in $\mathbb{R}_+^{m+1}$; here $x \in \mathbb{R}^m$, and $y > 0$. Sometimes we write x_0 for y.* The space $L_1^p(\mathbb{R}_+^{m+1})$ consists of those $F \in L^p(\mathbb{R}_+^{m+1})$ such that $\frac{\partial F}{\partial x_j} \in L^p(\mathbb{R}_+^{m+1})$, $j=0,1,\ldots,n$, with the obvious norm (the measure used is the usual Lebesgue measure for $\mathbb{R}_+^{m+1}$). The connection with the Besov space $B^p(\mathbb{R}^m)$ is given by Gagliardo's lemma:

---

*In the previous chapters we have used ρ instead of y, but the present notation is more convenient for our purposes here.

12.4 Lemma. *Suppose* F *is smooth in* $\overline{\mathbb{R}}^{m+1}_+$, *and belongs to* $L^p_1(\mathbb{R}^{m+1}_+)$ *where* $1 < p < \infty$. *Let* f *be the restriction of* F *to* $\mathbb{R}^m$. *Then*

$$f \in B^p(\mathbb{R}^m), \text{ and}$$

$$\|f\|_{B^p} \leq C\|F\|_{L^p_1}.$$

(See 4.3 in Chapter VI of [33]. This is the special case when $\alpha = 1$.)

The converse of this lemma is also true, but we shall need a generalization in terms of Poisson operators which we shall now consider.

d. Poisson operators

We consider mappings of function f in $\mathbb{R}^m$ to function F on $\mathbb{R}^{m+1}_+$, given by

$$F(x,y) = \frac{1}{(2\pi)^m} \int_{\mathbb{R}^m} p(x,y,\xi)\tilde{f}(\xi)\, e^{i\langle x,\xi\rangle}\, d\xi = P(f). \tag{12.5}$$

Here p is the "symbol" of the operator P as defined in (7.32) and (7.31). In the rest of this chapter and in the next, we shall limit ourselves to operators of Poisson type of order 0 (unless the contrary is stated).

12.6 Main lemma. *The operator* $f \longrightarrow P(f) = F$ *maps* $B^p(\mathbb{R}^n)$ *to* $L^p_1(\mathbb{R}^{n+1}_+)$.

To prove this we need the following:

12.7 Lemma. *The operator* $f \longrightarrow Q(f) = \dfrac{1}{(2\pi)^m} \displaystyle\int_{\mathbb{R}^m} q(x,\xi)\hat{f}(\xi)e^{-i\langle x,\xi\rangle}d\xi$,

<u>mapping functions on</u> $\mathbb{R}^m$ <u>to functions on</u> $\mathbb{R}^m$, <u>(where</u> $q(x,\xi)$ <u>belongs to the standard symbol class</u> $S^0_{1,0}$) <u>maps</u> $L^p(\mathbb{R}^m)$ <u>to</u> $L^p(\mathbb{R}^m)$, $1 < p < \infty$.

<u>Proof of Lemma 12.6</u>. We assume that $q(x,\xi)$ has compact support in x. Then by the Fourier Transform $q(x,\xi) = \frac{1}{(2\pi)^m} \int_{\mathbb{R}^m} \hat{q}(\lambda,\xi) e^{i\langle x,\lambda\rangle} d\lambda$ where

$$\sup_{\xi \in \mathbb{R}^n} \left| \left(\frac{\partial}{\partial\xi}\right)^\alpha \hat{q}(\lambda,\xi)\right| \le A_\alpha (1+|\xi|)^{-|\alpha|} (1+|\lambda|)^{-N} \tag{12.8}$$

for each $N > 0$.

However,

$$Q(f) = \frac{1}{(2\pi)^m} \int_{\mathbb{R}^m} q(x,\xi) \hat{f}(\xi) e^{i\langle x,\xi\rangle} d\xi = \frac{1}{(2\pi)^{2m}} \int_{\mathbb{R}^m} \left(\int_{\mathbb{R}^m} \hat{q}(\lambda,\xi)\hat{f}(\xi) e^{i\langle x,\xi\rangle} d\xi\right) e^{i\langle x,\lambda\rangle}$$

However, by the Marcinkiewicz multiplier theorem (see [33], Chapter 4) each of the operators

$$f \to \frac{1}{(2\pi)^m} \int_{\mathbb{R}^m} \hat{q}(\lambda,\xi) \hat{f}(\xi) e^{i\langle x,\xi\rangle} d\xi$$

is bounded on $L^p(\mathbb{R}^m)$, with norm $\le A(1+|\lambda|)^{-N}$. An integration in λ then proves Lemma 12.7.

To prove the main lemma, we prove first that $F \in L^p(\mathbb{R}^{n+1})$. Now $F = F(x,y)$, and for fixed y, $f \to F(x,y)$ is given by a pseudo-differential operator with symbol in the class $S^0_{1,0}$, uniformly in $y > 0$. Thus by Lemma 12.7

$$\int_{\mathbb{R}^m} |F(x,y)|^p dx \le A \int_{\mathbb{R}^m} |f(x)|^p dx, \qquad y > 0$$

and since p and thus F has finite support in y, then an integration in y gives

$$\int_{\mathbb{R}_+^{m+1}} |F(x,y)|^p dx\,dy \leq A' \int_{\mathbb{R}^m} |f(x)|\, dx\ .$$

Next consider $\frac{\partial}{\partial y} F(x,y)$. It will make estimates easier if we assume (as we will from now on) that $p(x,y,\xi)$ vanishes when $|\xi| \leq 1$, (because the error, given by symbols with compact support in ξ, clearly gives $F \in C_0^\infty(\overline{\mathbb{R}}_+^{m+1})$, whatever f we start with).

Now $\frac{\partial}{\partial y} F(x,y) = \frac{\partial}{\partial y} P(f) = \frac{\partial}{\partial y} P(f-f_\epsilon) + \frac{\partial}{\partial y} P(f_\epsilon)$. $\frac{\partial P}{\partial y}$ has symbol $\frac{\partial p}{\partial y}(x,y,\xi)$, and as one can observe $y\frac{\partial}{\partial y} p(x,y,\xi)$ is again a symbol of Poisson type 0, hence gives a bounded operator on L^p for each fixed y, uniformly in y, by Lemma 12.7. Thus,

$$\text{(12.9)} \qquad \int_{\mathbb{R}^m} |\frac{\partial P}{\partial y}(f-f_\epsilon)|^p (x,y)\,dx \leq Ay^{-p} \|f-f_\epsilon\|_p^p\ .$$

Now $\frac{\partial p}{\partial y}(x,y,\xi) = \sum_{j=1}^m \frac{\partial p}{\partial y}(x,y,\xi)\frac{\xi_j}{|\xi|^2} \cdot \xi_j$ and each $\frac{\partial p}{\partial y}(x,y,\xi)\frac{\xi_j}{|\xi|^2}$ again belongs to the class of symbols of the Poisson type 0. Thus invoking Lemma 12.7 again gives

$$\text{(12.10)} \qquad \int_{\mathbb{R}^m} |\frac{\partial P}{\partial y}(f_\epsilon)|^p (x,y)\,dy \leq A \sum_{j=1}^m \|\frac{\partial f_\epsilon}{\partial x_j}\|_p^p\ .$$

Now in (12.9) and (12.10) let $\epsilon = y$, if $0 < y \leq 1$ and integrate in y. This leads to

$$\int_{\mathbb{R}^m}\int_0^1 \left|\frac{\partial F}{\partial y}(x,y)\right|^p dy\,dx \le A\int_0^1 \epsilon^{-p}\|f-f_\epsilon\|_p^p\,d\epsilon$$

$$+ A\int_0^1 \|\nabla f_\epsilon\|_p^p\,d\epsilon$$

and so $\frac{\partial F}{\partial y} \in L^p(\mathbb{R}_+^{m+1})$, with norm bounded by $\|f\|_{B^p}$, by Lemma 12.1.

The consideration of $\frac{\partial F}{\partial x_j}$ $j=1,\ldots,m$ is analogous. In fact $\frac{\partial F}{\partial x_j}$ corresponds to "symbol" $\frac{\partial p(x,y,\xi)}{\partial x_j} + i\xi_j\, p(x,y,\xi)$, while, as we have observed $\frac{\partial p}{\partial x_j}$ and $y\, p(x,y,\xi)\xi_j$ are symbols of Poisson type, and hence for each fixed y give bounded operators on L^p, uniformly in j. This concludes the proof of the main lemma.

Chapter 13. The spaces $\Lambda_\alpha(\mathbb{R}^m)$ and $\Lambda_\alpha(\mathbb{R}_+^{m+1})$

We let $\Lambda_\alpha(\mathbb{R}^m)$ denote the standard Lipschitz spaces, $\alpha > 0$, as described in Chapter V, §4 of [33]. Thus a bounded function f belongs to Λ_α, where $0 < \alpha < 1$, when $\|f(x-t) - f(x)\|_\infty \leq A|t|^\alpha$. (Here $\|\cdot\|_\infty$ denotes the sup norm.) For $\alpha = 1$, we require $\|f(x+t) + f(x-t) - 2f(x)\|_\infty \leq A|t|$, and when $\alpha > 1$, we proceed inductively, i.e., $f \in \Lambda_\alpha(\mathbb{R}^m) \Longleftrightarrow f \in L^\infty$ and $\frac{\partial f}{\partial x_j} \in \Lambda_{\alpha-1}(\mathbb{R}^m)$, $j=1,\ldots,m$.

On $\mathbb{R}_+^{m+1}$ we define the space $\Lambda_\alpha(\overline{\mathbb{R}}_+^{m+1})$ to consist of all functions on $\overline{\mathbb{R}}_+^{m+1}$ which can be extended to $\mathbb{R}^{m+1}$ so as to belong to $\Lambda_\alpha(\mathbb{R}^{m+1})$. The norm is the quotient norm. More precisely, let η be the closed subspace of $\Lambda_\alpha(\mathbb{R}^{m+1})$ consisting of all $F(x,y)$ $(x = (x_1 \ldots x_m), y = x_0)$, so that $F(x,y) = 0$ in $\overline{\mathbb{R}}_+^{m+1}$. Define $\Lambda_\alpha(\overline{\mathbb{R}}_+^{m+1}) = \Lambda_\alpha(\mathbb{R}^{m+1})/\eta$.

a. A characterization

In analogy with what was done for B^p, we characterize $\Lambda_\alpha(\mathbb{R}^m)$ in terms of rate of approximations. We shall state the analogue of Lemma (12.1) somewhat differently.

13.1 Lemma. Suppose $0 < \alpha < 1$. Then $f \in \Lambda_\alpha(\mathbb{R}^m)$ if and only if we can write

$$f = \sum_{k=0}^{\infty} g_k \tag{13.2}$$

where $\|g_k\|_{L^\infty} \leq A\,2^{-k\alpha}$ and

$$\|\nabla g_k\|_{L^\infty} \leq A\,2^{k-k\alpha}.$$

Proof. This lemma is not new. To prove it, let $u(x,y)$ be the Poisson integral of f (as in Chapter 5 of [33]) and use the fact that

$$(13.3)\qquad \left\|\frac{\partial u(x,y)}{\partial y}\right\|_{L^\infty} \le A\,y^{-1+\alpha} \quad\text{and}\quad \left\|\frac{\partial u}{\partial x_j}(x,y)\right\|_{L^\infty} \le A\,y^{-1+\alpha}$$

(see (49) and (51) of that reference), and set

$$g_k(x) = u(x, 2^{-k}) - u(x, 2^{-k+1})$$

$$g_0(x) = u(x, 1).$$

Now $g_k(x) = -\int_{2^{-k}}^{2^{-k+1}} \frac{\partial u}{\partial y}\,dy$, so $\|g_k\|_{L^\infty} \le A\,2^{-k\alpha}$;

while $\|\nabla g_k\| \le A\,2^{k-k\alpha}$, by the second inequality in (13.3).

Conversely, suppose f can be written in the form (13.2). Then

$$f(x-t) - f(x) = \Delta_t(f)$$

$$= \sum_{k=0}^{N} \Delta_t(g_k) + \sum_{N+1}^{\infty} \Delta_t(g_k)\,.$$

For the second sum we use the estimate $\|\Delta_t(g_k)\|_\infty \le 2\,\|g_k\|_\infty \le A\,2^{-k\alpha}$.

Thus $\left\|\sum_{N+1}^{\infty} \Delta_t(g_k)\right\| \le 2^{-N\alpha}$.

For the first sum use the estimate $\|\Delta_t(g_k)\|_{L^\infty} \le |t|\,\|\nabla g_k\|_{L^\infty} \le A|t|\,2^{k-k\alpha}$

Thus $\left\|\sum_{k=0}^{N} \Delta_t(g_k)\right\|_\infty \le A|t|\,2^{N-N\alpha}$.

Now if $|t| \le 1$, set N so that $2^{-N} \approx |t|$ and the result is $\|\Delta_t f\|_\infty \le A|t|^\alpha$, proving the lemma.

There is an analogue of the lemma which holds for $0 < \alpha < 2$. Again f is to have the decomposition (13.2) and

$$\begin{cases} \|g_k\|_{L^\infty} \le A\, 2^{-k\alpha} \\ \|\nabla g_k\|_{L^\infty} \le A\, 2^{k-k\alpha} \\ \|\nabla^2 g_k\|_{L^\infty} \le A\, 2^{2k-k\alpha} \end{cases} \tag{13.4}$$

($\|\nabla^2 g\|$ means the norm of all second derivatives.)

b. Λ_α and pseudo-differential operators

Suppose $Q(f) = \frac{1}{(2\pi)^m} \int_{\mathbb{R}^m} q(x,\xi) \hat{f}(\xi) e^{i\langle x,\xi\rangle} d\xi$ is a standard pseudo-differential operator, with $q \in S^0_{1,0}$.

(13.5) Lemma. <u>Q is a bounded operator of $\Lambda_\alpha(\mathbb{R}^m)$ to itself,</u> $\alpha > 0$.

Proof. By the usual commutation properties of Q and $\frac{\partial}{\partial x_j}$, one can easily reduce to the case $0 < \alpha \le 1$. We shall consider in detail this case.

We need a preliminary lemma. Suppose $\eta(\xi)$ is a fixed C^∞ function which vanishes when $|\xi| \le \frac{1}{4}$, or $|\xi| \ge 4$.

Let Q_k denote the pseudo-differential operator whose symbol is $q(x,\xi)\eta(\xi 2^{-k})$.

13.6 Lemma. <u>One has the following estimates:</u>

$$\left\| \left(\frac{\partial}{\partial x}\right)^\beta Q_k(f) \right\|_{L^\infty} \le A_\beta 2^{k|\beta|} \|f\|_{L^\infty}, \quad \text{for each } \beta. \tag{13.7}$$

Proof of Lemma 13.6. Suppose for simplicity that $\beta = 0$. Let

$$K_k(x, z) = \int_{\mathbb{R}^m} q(x, \xi)\, \eta(\xi 2^{-k})\, e^{i\langle z, \xi\rangle}\, d\xi .$$

Then $Q_k(f)(x) = \int K_k(x, z)\, f(x-z)\, dz$ and it suffices to see that

$$(13.8) \qquad \sup_{x \in \mathbb{R}^m} \int_{\mathbb{R}^m} |K_k(x, z)|\, dz \le A . \qquad \text{(with A independent of k)}$$

Now since

$(-iz)^\alpha K_k(x, z) = \int \left(\frac{\partial}{\partial \xi}\right)^\alpha (q(x, \xi)\, \eta(\xi 2^{-k}))\, e^{i\langle z, \xi\rangle}\, d\xi$ and $\eta(\xi 2^{-k})$ is non-vanishing only when $2^{k-2} \le |\xi| \le 2^{k+2}$, obvious estimates show that

$$(13.9) \qquad |z|^{|\alpha|}\, |K_k(x, z)| \le A_\alpha\, 2^{-k|\alpha| + mk} .$$

Now

$$\int_{\mathbb{R}^m} |K_k(x, z)|\, dz = \int_{|z| \le 2^{-k}} + \int_{|z| \ge 2^{-k}}$$

For the first integral use the estimate $|K_k(x, z)| \le A\, 2^{mk}$ (the case $\alpha = 0$ of (13.9)).

For the second integral use the estimate $|K_k(x, z)| \le A |z|^{-m-1} 2^{-k}$, (the estimate $|\alpha| = m+1$ in (13.9)). The result is (13.8), proving (13.7) with $\beta = 0$. The cases corresponding to $\beta \ne 0$ are done in the same way, concluding the proof of Lemma (13.6).

To prove Lemma (13.5), we need for an $f \in \Lambda_\alpha$ a decomposition of the type (13.2), where, however, the elements g_k have their spectrum contained in $2^{k-1} \le |\xi| \le 2^{k+1}$, for $k=1, 2, \dots,$.

To do this fix a ψ which is an even function in ξ, C^{∞}, and $\psi(\xi) = 1$ for $|\xi| \leq 1$, with $\varphi(\xi) = 0$ for $|\xi| \geq 2$. Now define g_k by

$\hat{g}_k(\xi) = [\psi(\xi 2^{-k}) - \psi(\xi 2^{-k+1})]\hat{f}(\xi), \quad k=1,2,\ldots,$

$\hat{g}_0(\xi) = \psi(\xi)\hat{f}(\xi).$

Then since $\psi(\xi 2^{-k}) \longrightarrow 1$ as $k \longrightarrow \infty$, we have $f = \sum_{k=0}^{\infty} g_k$. Moreover, $g_k = \varphi_k * f$, $k=1,\ldots,$ where $\varphi_k(x) = 2^{mk}\varphi(2^k x)$, with $\hat{\varphi}(\xi) = \psi(\xi) - \psi(2\xi)$. Observe that $\psi(\xi 2^{-k}) - \psi(\xi 2^{-k+1})$ is supported in $2^{k-1} \leq |\xi| \leq 2^{k+1}$. Also

$$g_k(x) = \int \varphi_k(t) f(x-t)\,dt = \int \varphi_k(t)[f(x-t) - f(x)]\,dt$$

$$= \frac{1}{2}\int \varphi_k(t)[f(x-t) + f(x+t) - 2f(x)]\,dt, \quad \text{since } \varphi_k \text{ is even and } \int \varphi_k(t)\,dt = 0.$$

Hence if $f \in \Lambda_\alpha$ $0 < \alpha \leq 1$, then

$$\|g_k(x)\|_{L^\infty} \leq A \int_{\mathbb{R}^m} |\varphi_k(t)|\, |t|^\alpha dt \cdot \|f\|_{\Lambda_\alpha} \quad \text{and}$$

$$\int_{\mathbb{R}^m} |\varphi_k(t)|\, |t|^\alpha dt = 2^{nk}\int_{\mathbb{R}^m} |\varphi(2^k t)|\, |t|^\alpha dt = 2^{-k\alpha} A.$$

This shows $\|g_k\|_{L^\infty} \leq A\, 2^{-k\alpha}$.

By the same arguments one also proves the full set of inequalities (13.4).

We can now finish the proof of Lemma 13.5. Suppose, therefore, $f \in \Lambda_\alpha$, $0 < \alpha \leq 1$. Now $Qf = \sum_{k=0}^{\infty} Qg_k$, and by Lemma 13.1 and the remarks that follow it suffices to see that

$$\text{(13.10)} \qquad \left\|\left(\frac{\partial}{\partial x}\right)^\beta Qg_k\right\|_{L^\infty} \leq A\, 2^{-k\alpha + |\beta| k}, \qquad 0 \leq |\beta| \leq 2.$$

The terms corresponding to $k=0$ can be disregarded since g_0 has spectrum contained in $|\xi| \leq 2$, and so Qg_0 is as regular as we wish. Now in the definition of Q_k (see the remarks preceding Lemma 13.6) take $\eta(\xi)$ so as to be $= 1$, when $\frac{1}{2} \leq |\xi| \leq 2$. Then

$$Q(g_k) = Q_k(g_k), \qquad k=1,2,\ldots, \tag{13.11}$$

since Q and Q_k agree on the spectrum of g_k. We now invoke (13.7) with $f = g_k$, and use (13.4), the result is (13.10) and Lemma 13.5 is proved

We come to our main result.

13.12 Main lemma. If P is an operator of Poisson-type of order 0 then P maps $\Lambda_\alpha(\mathbb{R}^m)$ to $\Lambda_\alpha(\mathbb{R}^{m+1}_+)$.

Proof. We consider first the case when $0 < \alpha \leq 1$. Write $F = P(f)$, and if

$$f = \sum_{k=0}^{\infty} g_k, \quad \text{then } G_k = P(g_k)$$

and

$$F = \sum_{k=0}^{\infty} G_k. \quad \text{We prove that}$$

$$\begin{cases} \|G_k\|_{L^\infty} \leq A\, 2^{-k\alpha} \\ \|\nabla G_k\|_{L^\infty} \leq A\, 2^{k-k\alpha} \\ \|\nabla^2 G_k\|_{L^\infty} \leq A\, 2^{2k-k\alpha} \end{cases} \tag{13.13}$$

where ∇ means the gradient $\frac{\partial}{\partial x_1}, \frac{\partial}{\partial x_2}, \ldots, \frac{\partial}{\partial y}$, and ∇^2 means the matrix of all second partial derivatives (including the $\frac{\partial}{\partial y}$'s). This is proved in the same way as in the proof of Lemma 13.5 before, when we observe (as we already have before) that $p(x,y,\xi)$ is type $S^0_{1,0}$ for each fixed y uniformly in y, and $\frac{\partial}{\partial y} p(x,y,\xi) \frac{\xi_j}{|\xi|^2}$, are also of type $S^0_{1,0}$, uniformly in y, etc.

Now (13.13) does not yet conclude the proof of the theorem (when $0 < \alpha \leq 1$). We need an extension operator, mapping functions on $\overline{\mathbb{R}}^{m+1}_+$ to functions on $\mathbb{R}^{m+1}$. We use the mapping $F \longrightarrow \widetilde{F}$

where $\widetilde{F}(x,y) = F(x,y)$ if $y > 0$

and

$$(13.14) \qquad \widetilde{F}(x,y) = \int_1^\infty F(x, y(1-2\lambda))\,\psi(\lambda)\,d\lambda\,, \quad \text{if } y < 0,$$

as already described in Chapter 7 (see (7.58) and the discussion that follows).

A basic property of this mapping is the fact that

$$(13.15) \qquad \|\widetilde{F}\|_{L^p_k(\mathbb{R}^{m+1})} \leq A_k \|F\|_{L^p_k(\mathbb{R}^{m+1}_+)}, \quad 1 \leq p \leq \infty,$$

and $k=0,1,2,\ldots$. For the proof of this see [33], Chapter VI, §3.

Now $\widetilde{F} = \sum \widetilde{G}_k$,

and by (13.15) the $\widetilde{G}_k$ satisfy the same kind of inequalities (i.e., (13.13)) as do the G_k, but now on all of $\mathbb{R}^{m+1}$.

Thus by Lemma (13.1), (and the variant alluded to at the end of its proof) $\widetilde{F} \in \Lambda_\alpha(\mathbb{R}^{m+1})$. Since the restriction of $\widetilde{F}$ to $\mathbb{R}^{m+1}_+$ is F, we get $F \in \Lambda_\alpha(\mathbb{R}^{m+1}_+)$ as desired. This proves the result when $0 < \alpha \leq 1$. For higher values of α we use recursion. Thus $F = P(f) \in \Lambda_\alpha(\mathbb{R}^{m+1}_+)$, with $\alpha > 1$, if the first partials of $F \in \Lambda_{\alpha-1}(\mathbb{R}^{m+1}_+)$. But these can be expressed as Poisson-type operators of the first partials of f, and these are in $\Lambda_{\alpha-1}(\mathbb{R}^m)$, etc.

Bibliographical comments for Chapters 12 and 13

1. In using the reference [33], the reader should be warned that throughout §5.1 and 5.2 of Chapter V there occurs a systematic slip in sign: The $\alpha-1$ in expressions such as (61), (62), should be replaced by $-\alpha+1$; also $\alpha-k$ by $-\alpha+k$, etc.

2. The one-dimensional version of Lemma 13.1 and (13.4) is in Zygmund [36], Chapter III, (13.14) and (13.20). Note that what we call Λ_1 he calls Λ_*.

3. Lemmas 12.1, 12.4, and 13.1 can be found in Nikol'skii [25], Chapters 4-6, in considerably more general form. He denotes B^p and Λ_α by $B^{1-1/p}_{p,p}$ and H^α_∞, respectively.

4. Probably the only novel results contained in these chapters are Main Lemmas 12.6 and 13.12 dealing with Poisson operators. For some related earlier estimates see Agmon, Douglis, and Nirenberg [0].

Chapter 14. The spaces B^p, L^p_1, Λ_α on M and bM

a. Definitions

We come closer to our ultimate applications.

M is a domain in a complex manifold with smooth boundary bM. What we have done above will make it easy to define the spaces $B^p(bM)$, $\Lambda_\alpha(bM)$; also $L^p_k(\overline{M})$ and $\Lambda_\alpha(\overline{M})$. In fact suppose, for example, $f \in B^p(\mathbb{R}^m)$. Then if φ is a local diffeomorphism of $\mathbb{R}^m$ and $\psi \in C_0^\infty$ whose support is contained where φ is regular, then $\psi(x)f(\varphi(x)) \in B^p(\mathbb{R}^m)$. This follows immediately for the characterization given in Lemma 12.1. This allows one to define $B^p(bM)$ in terms of a finite patching of coordinate neighborhoods of bM. Similarly one can define the spaces $\Lambda_\alpha(bM)$; first by using Lemma 13.1(see also 13.4)), when $0 < \alpha \leq 1$; and then for higher α recursively by differentiation.

The space $L^p_k(\overline{M})$ has a standard definition which needs no further comment. The space $\Lambda_\alpha(\overline{M})$ can also be defined. It needs two kinds of patchings: The first in terms of interior neighborhoods, which reduce matters to $\Lambda_\alpha(\mathbb{R}^{m+1})$; the second in terms of boundary neighborhoods which reduce matters to $\Lambda_\alpha(\overline{\mathbb{R}}^{m+1}_+)$. Again nothing new is really involved.

b. Four types of operators

We shall describe four types of operators in terms of which our estimates will be stated.

(i) Operators of type I. There is only a single operator in this class It is the restriction operator mapping functions on $\overline{M}$ to functions on bM by $$f \longrightarrow f|_{bM}.$$

(ii) Operators of type II. These are the operators mapping functions on bM to functions on bM, which in coordinate neighborhoods are given by classical pseudo-differential operators of order zero, (i.e., with symbol in $S^0_{1,0}$).

(iii) Operators of type III. These are operators mapping functions on bM to functions on bM, which are given by Heisenberg-group type kernels of type λ, $\lambda = 0$, $\lambda = 1$ or $\lambda = 2$, (see [9], p. 486). The operators will also be written as III_0, III_1, and III_2, to indicate the λ.

(iv) Operators of type IV. These are operators mapping functions on bM to functions on $\overline{M}$, i.e., $f \longrightarrow F$. F is always C^∞ in the interior of $\overline{M}$ (no matter what f is). For points near the boundary bM, these operators are given in appropriate coordinate neighborhoods by operators of Poisson-type (order 0), as in (12.5), and (7 31).

c. The estimates

All mappings will be bounded in the norm. Also $1 < p < \infty$; and $\alpha > 0$.

14.1 Theorem I. The operator of type I maps

(a) $L^p_1(\overline{M})$ to $B^p(bM)$

(b) $\Lambda_\alpha(\overline{M})$ to $\Lambda_\alpha(bM)$

14.2 Theorem II. Operators of type II map

(a) $B^p(bM)$ to $B^p(bM)$

(b) $\Lambda_\alpha(bM)$ to $\Lambda_\alpha(bM)$

14.3 Theorem III. Operators of type III map

(a) $B^p(bM)$ to $B^p(bM)$, when the operator is of type 0.

(b) $\Lambda_\alpha(bM)$ to $\Lambda_{\alpha+\lambda/2}(bM)$, and $L^\infty(bM)$ to $\Lambda_{\lambda/2}(bM)$, when the operator is of type λ, $\lambda = 1,2$.

14.4 Theorem IV. Operators of type IV map

(a) $B^p(bM)$ to $L_1^p(\overline{M})$

(b) $\Lambda_\alpha(bM)$ to $\Lambda_\alpha(\overline{M})$

Proof of Theorem I. Part (a) follows from Lemma 12.4, Lemma 12.1 which is the characterization in terms of approximations, and the local definition of $B^p(bM)$. Part (b) needs no comment.

Proof of Theorem II. Part (a): according to Lemma 12.7, an operator of type II maps $L^p(bM)$ to itself boundedly. Such operators commute with vector fields (modulo operators again of type II) and so by the characterization Lemma 12.1, we get part (a). Part (b) is proved similarly, invoking Lemma 13.5, and the characterization Lemma 13.1 (or (13.4)).

Proof of Theorem III. Part (a). We need the following facts.

(i) Each operator of type III_0 is bounded on $L^p(bM)$ to itself. See [9], §15.

(ii) If $X_1, \ldots, X_m$ is a smooth basis of vector fields and S is any operator of type III_0, then $\exists\, S_0^{(j)}, S_1^{(j)}, \dots, S_m^{(j)}$, all of type III_0, so that

$$(14.5) \qquad X_j S = \sum_{k=1}^{m} S_k^{(j)} X_k + S_0^{(j)} .$$

This is (implicit) in [9], p. 490; also see Rothschild-Stein [29].

Since $f \in B^p(bM)$ can be characterized in terms of rapidity of L^p approximation, with approximating function controlled in $L_1^p(bM)$ norms, part (a) then follows from facts (i) and (ii).

Part (b). This part we announced in [12], Lemma (6.2), part (b). The idea we had in mind for the proof of the lemma was later generalized, and in a more general form appears in Rothschild-Stein [29], Theorem 14. There seems little point in repeating that argument here.

Proof of Theorem IV. Part (a) and (b) are simply consequences of Main Lemma 12.6, and Main Lemma 13.3, together with the remarks already made.

Chapter 15. Main results

Let N denote the (exact) Neumann operator described in Theorem 11.3. Our purpose here will be to prove the regularity of N in terms of function spaces $L^p_k(\overline{M})$, $\Lambda_\alpha(\overline{M})$, and others that will be defined below. For simplicity of notation we are using $L^p_k(\overline{M})$ to denote not only the previously defined space of scalar-valued functions, but also its analogue of (0,1) forms on M whose components belong to $L^p_k(\overline{M})$; similarly for the other spaces studied in Chapter 14. But this abuse of notation should not lead to any confusion. In all our theorems we have $1 < p < \infty$.

We shall say that a smooth vector field X defined on $\overline{M}$ is allowable if restricted to the boundary it points in the "good" directions, i.e., $X|_{bM} \in T^{1,0}(bM) \oplus T^{0,1}(bM)$.

15.1 Theorem. N has a unique extension so that the indicated mappings are bounded.

(a) $N: L^p_k(\overline{M}) \longrightarrow L^p_{k+1}(\overline{M})$, $1 < p < \infty$; $k=0,1,2,\ldots$

(b) $p(Z,\overline{Z})N: L^p_k(\overline{M}) \longrightarrow L^p_k(\overline{M})$, $k=0,1,2,\ldots,$

where p is any second-degree polynomial in allowable vector fields.

(c) $\overline{Z}_{n+1}N: L^p_k(\overline{M}) \longrightarrow L^p_{k+1}(\overline{M})$, $k=0,1,2,\ldots$ *

---

*In any appropriate coordinate neighborhoods, the vector fields Z_j, $j=1,\ldots,n$ give a basis for the holomorphic vector fields which are tangential at bM. Thus Z_j and $\overline{Z}_j$ are allowable $1 \leq j \leq n$. However, Z_{n+1} is the holomorphic vector field which (near bM) has the property that $\text{Re}\, Z_{n+1} = 1/\sqrt{2}\ \partial/\partial\rho$, where ρ is the geodesic distance from bM. The exact form of Z_{n+1} away from bM is of no significance.

We shall prove these results by showing the boundedness of the mappings on $C^\infty(\overline{M})$. The rest then follows by a simple limiting argument. Now if $f \in C^\infty(\overline{M})$, then $u = N(f) \in C^\infty(\overline{M})$, and u solves the $\bar{\partial}$-Neumann problem (11.1) and (11.2).* For u we apply Proposition 9.26, when $n > 1$, or Proposition 10.30 when $n = 1$ to express (in appropriate coordinate patches) u in terms of our approximate Neumann operator N_a, together with the remainder $R(u)$. The question then becomes that of estimating N_a and R.

Proof of part (a). The first task will be to prove:

15.2 Proposition. N_a is bounded from L^p_k to L^p_{k+1}, $k=0,1,2\ldots$

We can write N_a symbolically (disregarding smooth cut-off functions) in the form

$$N_a = G + PS_2T_1 \operatorname{Rest} D_1 G \tag{15.3}$$

where G is the Green's operator which by the elliptic theory is smoothing of order 2 in all directions; D_1 is a differential operator of order 1; Rest is the restriction-to-the-boundary operator (type I in the terminology of Chapter 14); T_1 is a standard pseudo-differential operator of order 1 on bM; S_2 is an operator of type III_2; and P is the Poisson operator, type IV). When $n = 1$ the description of N_a is slightly more complicated, and we shall discuss this case separately below.

The proof of Proposition 15.2 requires a lemma.

15.4 Lemma. G is a bounded mapping from L^p_k to L^p_{k+2}, $k=0,1,2,\ldots.$

*See Note on page 133.

It suffices to prove this in the separate coordinate neighborhoods with which we are dealing. Then the problem is reduced to corresponding problem in $\mathbb{R}_+^{m+1} = \mathbb{R}_+^{2n+2}$. Let us recall the extension operator $\sim$ (used in (7.58)), and also in the proof of Lemma 13.12. By (13.15) we know that it is a bounded operator from $L_k^p(\mathbb{R}_+^{m+1})$ to $L_k^p(\mathbb{R}^{m+1})$. Now (see (7.63))

$$(15.5) \qquad G(f) = E(\tilde{f}) - P((E\tilde{f})_0)$$

E is a standard pseudo-differential operator of order -2, therefore by "commuting" derivatives past E, we see by Lemma (12.7) that E maps $L_k^p(\mathbb{R}^{m+1})$ to $L_{k+2}^p(\mathbb{R}^{m+1})$. Next we consider the term $P((E\tilde{f})_0)$. We observe by the symbolic calculus that

$$(15.6) \qquad \frac{\partial}{\partial x_j} P = P \frac{\partial}{\partial x_j} + P'$$

and

$$(15.7) \qquad \frac{\partial}{\partial y} P = \sum_j P_j \frac{\partial}{\partial x_j} + P''$$

where P_j, P' and P'' are Poisson operators of order 0.

In taking a derivative of order $k+2$ of $P((E\tilde{f})_0)$, we can commute $k+1$ of the derivatives past P, and then past E; turning E into pseudo-differential operator of order 0, and having $k-1$ derivatives act on on $\tilde{f}$. This results in a function in $L_1^p(\mathbb{R}^{m+1})$, and the restriction gives an element in B^p. Finally an application of Lemma 12.6 concludes the proof of the present lemma.

For further reference we record the Λ_α analogue of the lemma, which is proved very similarly.

(15.8) Lemma. G is a bounded mapping from $\Lambda_\alpha(\mathbb{R}^{m+1}_+)$ to $\Lambda_{\alpha+2}(\mathbb{R}^{m+1}_+)$, $\alpha > 0$.

We return to the proof of Proposition 15.2, and now we need to concern ourselves only with the second term of the right side of (15.3). If we apply k derivatives we obtain (after using (15.6) and (15.7)) a sum of sums of the form

$$\Sigma P_\alpha \left(\frac{\partial}{\partial x}\right)^\alpha S_2\left(\frac{\partial}{\partial x_j}\right) T_0 \operatorname{Rest} D_1 G(f), \quad \text{where } |\alpha| \leq k, \tag{15.9}$$

and where P_α are each Poisson operators of order 0.

Now by [9], Proposition 15.14, the operators $S_2 \frac{\partial}{\partial x_j}$ are each operators of the kind S_0 (type III_0 in our terminology). Because of (14.5) (or more precisely its transpose) we can "commute" the operators $\left(\frac{\partial}{\partial x}\right)^\alpha$ past $S_2 \frac{\partial}{\partial x_j}$, and then combine them with D_1. The result replaces each sum in (15.9) by sums of terms of the form

$$\Sigma P_\alpha S_0 \operatorname{Rest} \left(\frac{\partial}{\partial x}\right)^\beta G(f), \qquad |\beta| \leq k+1. \tag{15.10}$$

Now $f \in L^p_k(\overline{\mathbb{R}}^{m+1}_+)$, thus by Lemma (15.4) $G(f) \in L^p_{k+2}$, and $\left(\frac{\partial}{\partial x}\right)^\beta G(f) \in L^p_1$. So in view of Lemma 12.4 $\operatorname{Rest}\left(\frac{\partial}{\partial x}\right)^\beta G(f) \in B^p$, and by Theorem 14.3 (a), the result of applying S_0 is in B^p. Finally by Theorem 14.4 (a), (15.10) belongs to L^p_1. Since this is any derivative of order k of the P term in (15.3), and since all our inclusions are bounded mappings, Proposition 15.2 is completely proved, when $n > 1$.

In the case $n = 1$, the modification required is that $S_2 T_1$ has to be replaced by $T_0S_2T_1 + T_{-1}$ (see Proposition 10.30 and (10.23)); but these terms can absorb any derivative of order 1, giving nothing worse than $T_0S_0T_1$, and the argument then is carried out as before. This proves Proposition 15.2. To handle the remainder term (see (9.25)) we need the following:

<u>(15.11) Lemma.</u> <u>Let</u> $1 < p < \infty$, $k=1,2,3,\ldots$.
<u>Then there exists</u> $k' < k$, <u>so that</u> R <u>is bounded from</u> $L^p_{k'}$ <u>to</u> L^p_k.

The difficulty with the operator R is that it is smoothing of at most order $\frac{1}{2}$. One could overcome this difficulty by an iteration argument, but a more serious obstacle would still stand in the way, namely that R cannot be bounded on L^p because restriction of an L^p function to a hyper-plane makes no sense. Thus we need to start with $L^p_{k'}$, where $k' > 0$, (in fact $k' > 1/p$).

The elements of the fractional Sobolev space $L^p_k(\mathbb{R}^{m+1}_+)$ are defined as restrictions of elements in $L^p_k(\mathbb{R}^{m+1})$, i.e., $L^p_k(\mathbb{R}^{m+1}_+) = L^p_k(\mathbb{R}^m)/\eta$, where η is the subspace of $L^p_k(\mathbb{R}^{m+1})$ consisting of those functions which vanish on $\mathbb{R}^{m+1}_+$. (For the facts about the spaces $L^p_k(\mathbb{R}^{m+1})$ see e.g., [33], Chapter V.) Because of the property (13.15) of the extension map this is consistent with the usual definition when k is an integer. Moreover because of (13.15), and the standard interpolation theorem for $L^p_k(\mathbb{R}^{m+1})$ (see [4], Theorem 10), it follows that whenever $f \in L^p_k(\mathbb{R}^{m+1}_+)$, then $\tilde{f} \in L^p_k(\mathbb{R}^{m+1})$ and the mapping is bounded, for any real k.

In proving the lemma we limit ourselves to the least regular term of R which can be written as

(15.12) $A(f) = P(S_1 \text{ Rest } T_0(\tilde{f}))$.

We observe first that:

(15.13) The mapping $f \longrightarrow (A(f))^{\sim}$ is bounded from L^2_k to L^2_1, if $k > 1/2$.

Suppose $f \in L^2_k$. First $\tilde{f} \in L^2_k$ as we have already pointed out, and therefore so is $T_0(\tilde{f})$. Secondly, Rest maps L^2_k to $L^2_{k-1/2}$ (if $k > 1/2$) (see [17], Chapter II, or [33], Chapter V), and this is included in L^2. Next an operator of type III_1 (i.e., S_1) maps L^2 to $L^2_{1/2}$ (see [9], Proposition 19.7 and [29], Lemma 16.1). But since $L^2_{1/2}(\mathbb{R}^m) = B^2(\mathbb{R}^m)$, (see [33], Chapter V, §3.5), Lemma 12.6 shows that P maps the result to L^2_1. Thus (15.13) is proved.

We also observe that

(15.14) The mapping $f \longrightarrow (A(f))^{\sim}$ is bounded from L^p_k to L^p_k, if k=1,2,

To prove this we can disregard the fact that S_1 has any smoothing properties, and since it is also an S_0, by Theorem 14.3 (a), it maps B^p to itself boundedly. Now if $f \in L^p_1$, then Rest $(T_0\tilde{f}) \in B^p$ and so $A(f)^{\sim} \in L^p_1$, by Theorem 14.4 (a). The case for $k > 1$ is proved similarly; (the argument here is very much the same as in the proof of Proposition 15.2).

Finally an application of the interpolation theorem for L^p_k spaces shows (see [4], Theorem 10) that (15.13) with (15.14) imply the required boundedness for A, and hence for R. Lemma 15.11 is therefore proved.

Now for appropriate cut-off functions φ we have $\varphi u - R(u) = N_a(f)$ (see Propositions 9.26 or 10.30). Taking a sum over a finite covering of $\overline{M}$ by coordinate patches, and invoking Lemma (15.11) and Proposition 15.2 gives

(15.15) $$\|u\|_{L^p_{k+1}} \leq A\{\|f\|_{L^p_k} + \|u\|_{L^p_{k'+1}}\} \text{ where } k' < k \text{ and } k=0,1,2,\ldots$$

Now it is well known that by holomorphic convexity arguments one gets $\|u\|_{L^p_a} \leq C\|u\|^{\theta}_{L^p} \|u\|^{1-\theta}_{L^p_b}$, where a and b are any positive numbers, $0 < a < b$, and $0 < \theta < 1$. (See Hirschmann [16] and Calderón [14].) From this it follows that for any $\varepsilon > 0$, there exists a constant C_ε so that

(15.16) $$\|u\|_{L^p_{k'+1}} \leq \varepsilon\|u\|_{L^p_{k+1}} + C_\varepsilon\|u\|_{L^p}, \quad \text{if } k' < k.$$

Inserting (15.16) in (15.15) (with ε chosen so that $A\varepsilon \leq 1/2$) leads to

(15.17) $$\|u\|_{L^p_{k+1}} \leq A'\{\|f\|_{L^p_k} + \|u\|_{L^p}\}.$$

We are now very close to our goal, and need only remove the term $\|u\|_{L^p}$ from the right side. This is done by proving

(15.18) $$\|u\|_{L^p} \leq A\|f\|_p, \qquad 1 < p < \infty.$$

By the L^2 theory (see Theorem 11.3, (iii))

(15.19) $$\|u\|_{L^2} \leq A\|f\|_{L^2}$$

Next if p is sufficiently large ($p > 2n+2$), then by Sobolev's theorem $\|u\|_{L^\infty} \leq A\|u\|_{L^p_1}$ (see [33], Chapter V, §2). Thus the case $k=0$ of (15.17) gives

(15.20) $$\|u\|_{L^\infty} \leq A' \{\|f\|_{L^p} + \|u\|_{L^p}\}$$

But if $p > 2$, then an elementary argument shows that

$$\|u\|_{L^p} \leq \epsilon \|u\|_{L^\infty} + C_\epsilon \|u\|_{L^2} .$$

Combining this with (15.19), the fact that $\|f\|_{L^2} \leq A\|f\|_{L^p}$ if $p \geq 2$, and (15.20) finally leads to

$$\|u\|_{L^\infty} \leq A\|f\|_{L^p}$$

and hence (15.18) if $p > 2n+2$.

Then a standard interpolation theorem, using (15.19), gives (15.18) for $2 \leq p < \infty$. Finally, since N is self-adjoint, a duality argument also proves (15.18) when $1 < p \leq 2$. With this and (15.17) we have proved that N restricted to $C^\infty(\overline{M})$ is bounded from $L^p_k(\overline{M})$ to $L^p_{k+1}(\overline{M})$, and hence has a unique bounded extension, proving part (a) of Theorem 15.1.

<u>Proof of part (b) of Theorem 15.1</u>. We have first

<u>(15.21) Proposition</u>. $p(Z,\overline{Z})N_a$ <u>is bounded mapping from</u> $L^p_k(\overline{M})$ <u>to</u> $L^p_k(\overline{M})$, $k=0,1,2,\ldots$.

The proof is very similar to that of Proposition 15.2 except now two differentiations in the "good" directions (i.e., in terms of Z_i and $\overline{Z}_j$, $1 \leq i, j \leq n$) can be absorbed in K giving an operator of type S_0, (III_0); (see [9], Proposition 15.14); the latter is bounded on B^p as we already know.

Next

$$(15.22)\qquad \|p(Z,\overline{Z})\,R(u)\|_{L^p_k} \le A\|u\|_{L^p_{k+1}}\ ,\quad k=0,1,2,\dots .$$

This follows because in the main term of R (see (15.12)) one of the derivatives in the "good" directions can be absorbed in S_1, giving an operator of type III_0. Then the proof of (15.14) shows that (15.22) follows. So Proposition 9.26 or 10.30 imply that

$$\|p(Z,\overline{Z})u\|_{L^p_k} \le A\{\|f\|_{L^p_k} + \|p(Z,\overline{Z})R(u)\|_{L^p_k}\}$$

However, by part (a) of the theorem we have already proved, and (15.22), we get

$$\|p(Z,\overline{Z})u\|_{L^p_k} \le A\|f\|_{L^p_k}$$

and part (b) is also proved.

Proof of part (c) of Theorem 15.1. This is the deepest part of the theorem and requires the most delicate analysis so far. In explaining this it will be good to review some of the ideas of the construction of the approximate Neumann operator N_a.

One main task in Chapters 8 and 9 was to find the approximate left inverse of the operator $\Box^+$. The required operator was $-K\Box^-$, and we had in fact (see Proposition 8.19 and (9.10))

$$(15.23)\qquad -K\Box^-\,\Box^+ = I - \rho,$$

where the error term is of the form $\rho = S_1T_0 + T_{-1}$.

From this we can obtain a better approximation to a left inverse, namely for any integer m,

(15.24) $$-K_L \Box^- \Box^+ = I - \rho^m,$$

where

$$K_L = (I + \rho + \dots + \rho^{m-1})K.$$

Notice that the form ρ^m is now smoothing of a high order (in our various senses), if m is large.

Suppose we use K_L instead of in K in our approximate Neumann operator. Then we shall obtain an identity of the kind

$$\varphi u - R'(u) = N'_a(u)$$

N'_a is of the form (omitting cut-off functions)

(15.25) $$N'_a = G + P K_L \Box^- \text{ Rest } D_1 G$$

and R' is of the form

(15.26) $$R' = \tilde{T}_{-1} + P T_{-1} \text{ Rest } \tilde{T}_0 + \text{error}$$

where the error is smoothing of high order.

For the applications below we shall also need an approximation of high degree to the right-inverse of $\Box^+$. Now analogously to (15.23) we have

$$-\Box^+ \Box^- K = I - \rho'$$

where $\rho' = T_0 S_1 + T_{-1}$.

Thus

(15.24′) $$-\Box^+ \Box^- K_R = I - (\rho')^m,$$

if $K_R = K(I + \rho' + \dots + (\rho')^{m-1})$.

Hence $-K_L\Box^-\Box^+\Box^-K_R = (I - \rho^m)\Box^-K_R = K_L\Box^-(I - (\rho')^m)$.

So $\Box^-K_R - K_L\Box^- = \rho^m\Box^-K_R - K_L\Box^-(\rho')^m$. Finally

(15.27) $$-\Box^+K_L\Box^- = I + \text{error},$$

where

error $= \Box^+\{\rho^m\Box^-K_R - K_L\Box^-(\rho')^m\} - (\rho')^m$, which is smoothing of high degree if m is large.

We can come now to the proof of part (c) of Theorem 15.1.

<u>(15.28) Proposition.</u> $\overline{Z}_{n+1}N'_a$ <u>is bounded from</u> L^p_k <u>to</u> L^p_{k+1} $k=0,1,2,\ldots$.

It suffices to consider the P term in (15.25), the G term being smoothing of order 2 in all directions. Now by Lemma 8.17 $\overline{Z}_{n+1}P = P\Box^+ + Q_0$, where Q_0 is an operator of Poisson type of order 0. Thus $\overline{Z}_{n+1}P(K_L\Box^-\ldots) = P(\Box^+K_L\Box^-\ldots) + Q_0(K_L\Box^-\ldots)$. The Q_0 term is handled as in Proposition 15.2, and gives a bounded operator from L^p_k to L^p_{k+1}. Next, $P(\Box^+K_L\Box^-\ldots) = -P(I\ldots) + P(\text{error}\ldots)$ by (15.27), where $P(\text{error}\ldots)$ is smoothing of high order (and so maps L^p_k to L^p_{k+1}). Finally what remains is $P(\text{Rest } D_1G)$ which by our previous arguments maps L^p_k to L^p_{k+1}. The proof of Proposition 15.28 is complete.

<u>15.29 Lemma.</u> $\overline{Z}_{n+1}R'$ <u>maps</u> L^p_{k+1} to L^p_{k+1}, $k=0,1,2,\ldots$.

This is now a straightforward consequence of (15.26), and requires no further discussion.

We can now finish the proof of part (c). We have because of (15.25), Proposition 15.28 and Lemma 15.29,

$$\|\bar{Z}_{n+1}u\|_{L^p_{k+1}} < A\{\|\bar{Z}_{n+1}N_a(f)\|_{L^p_{k+1}} + \|\bar{Z}_{n+1}R(u)\|_{L^p_{k+1}} + \|u\|_{L^p_{k+1}}\}$$

$$\leq A\{\|f\|_{L^p_k} + \|u\|_{L^p_{k+1}}\} \leq A\|f\|_{L^p_k}$$

the last inequality by part (a) of the theorem. Thus Theorem 15.1 is now completely proved.

We now give the Λ_α analogue of Theorem 15.1, parts (a) and (c). A variant of part (b) in this context will be given below in Theorem 15.33.

<u>15.30 Theorem.</u>

(a) N <u>is bounded from</u> Λ_α <u>to</u> $\Lambda_{\alpha+1}$, $\alpha > 0$

(b) $\bar{Z}_{n+1}N$ <u>is bounded from</u> Λ_α <u>to</u> $\Lambda_{\alpha+1}$, $\alpha > 0$

<u>Proof.</u> Following the arguments for the L^p inequalities closely one can show that $N_a: \Lambda_\alpha \longrightarrow \Lambda_{\alpha+1}$, and $R: \Lambda_\alpha \longrightarrow \Lambda_{\alpha+1/2}$. (This because the terms S_j which occur in N_a and R map Λ_α to $\Lambda_{\alpha+j/2}$, by Theorem 14.3(b).) Thus as before

$$\|u\|_{\Lambda_{\alpha+1}} \leq A\{\|f\|_{\Lambda_\alpha} + \|u\|_{\Lambda_{\alpha+1/2}}\}.$$

As a consequence

$$(15.31)\qquad \|u\|_{\Lambda_{\alpha+1}} \leq A'\{\|f\|_{\Lambda_\alpha} + \|u\|_{\Lambda_\beta}\}, \quad \text{for } 0 < \beta < 1$$

However, by Theorem 15.1 $\|u\|_{L^p_1} \leq A\|f\|_{L^p} \leq A\|f\|_{\Lambda_\alpha}$.

Moreover if $p > 2n+2$, then a classical variant of Sobolev's theorem (see [25], Chapter 6) shows that $\Lambda_\beta \subset L^p_1$, where $\beta = 1 - \frac{2n+2}{p}$. Inserting this in (15.31) gives part (a) of the theorem. Part (b) is proved in the same way as part (c) of Theorem 15.1, using the refined approximate Neumann operator N'_a and its error term R'. The details may be left to the reader.

A slight modification of the argument proves

15.32 Corollary. Suppose $f \in L^\infty(\overline{M})$. Then

$$N(f) \in \Lambda_1 \text{ and } \overline{Z}_{n+1} N(f) \in \Lambda_1 .$$

For the proof we need to observe that if T_{-1} is a standard pseudo-differential operator of order -1 then $T_{-1}: L^\infty \longrightarrow \Lambda_1$. In fact $T_{-1} = T_{-1} J_{-1} J_1$, where J_α is the Bessel potential of order α; see [33], Chapter V. J_α is a pseudo-differential operator with symbol in the class $S^{-\alpha}_{1,0}$. Thus $T_{-1} J_{-1}$ has order 0, and J_1 maps L^∞ to Λ_1, (see e.g. [4], Theorem 8). Thus the assertion is proved if we appeal to Theorem 14.2 (b). Similarly, T_{-2} maps L^∞ to Λ_2.

The results on Lipschitz spaces for the Neumann operator have not been completely parallel with those for the Sobolev spaces, because we have not shown the full improvement that comes about in the "good directions" as in part (b) of Theorem 15.1. This we remedy now.

We shall first need some definitions; we begin by recalling one that we have already used.

A smooth vector field X defined on $\overline{M}$ is said to be allowable if its restriction to bM points in the "good directions," i.e., if $X|_{bM} \in T^{1,0}(bM) \oplus T^{0,1}(bM)$.

We also define the space $\Gamma_\alpha(\overline{M})$, in analogy with the space $\Gamma_\alpha(bM)$ studied in [9], p. 492. First if $0 < \alpha \leq 1$, then $f \in \Gamma_\alpha(\overline{M})$ if whenever $t \longrightarrow \gamma(t)$, $t \in [a,b]$ is an integral curve in $\overline{M}$ of an allowable vector field, the function $t \longrightarrow f(\gamma(t))$ is in the classical Λ_α space, as a function of t. In general, if $k < \alpha \leq k+1$, with k an integer, then $f \in \Gamma_\alpha(\overline{M})$ if $X_i \cdots X_j f \in \Gamma_{\alpha-j}(\overline{M})$, $j \leq k$ as $X_1, \ldots, X_j$ range over allowable vector fields.

15.33 Theorem. Suppose $f \in \Lambda_\alpha(\overline{M})$, $\alpha > 0$,

then $\quad N(f) \in \Gamma_{\alpha+2}(\overline{M}) \cap \Lambda_{\alpha+1}(\overline{M})$

Remark. The technical difficulty involved in dealing with the spaces Γ_α is that an operator of type S_0 does not preserve Λ_α, nor does an operator of type T_0 preserve Γ_α. This requires that we give an argument which is different from that of part (b) of Theorem 15.1.

The proof will require the following lemma.

15.34 Lemma. The Poisson operator described in coordinate patches by (7.52)-(7.56)) maps $\Gamma_{\alpha+1}(bM) \cap \Lambda_\alpha(bM)$ to $\Gamma_{\alpha+1}(\overline{M})$ if $\alpha > 0$, and $\Gamma_\alpha(bM)$ to $\Gamma_\alpha(\overline{M})$ if $0 < \alpha \leq 1$.

Proof. We notice first that restricting ourselves to a suitable coordinate patch, $P = P_0 + P_1$ where P_0 is bounded from $L^\infty(\mathbb{R}^m)$ to $L^\infty(\mathbb{R}^{m+1}_+)$ $(m=2n+1)$, and P_1 has the property that $\frac{\partial}{\partial y}P_1$, $\frac{\partial}{\partial x_j}P_1$, $P_1\frac{\partial}{\partial x_j}$ are all operators of Poisson type of order 0.

In fact by (7.55) P_0 has symbol $e^{-y\Delta}$ ($y=\rho$ in our terminology), with $\Delta^2 = \tau^2 + 2\sum_j |\sigma(Z_j)|^2 = Q_x(\tau, \xi) = Q_x(\xi')$, where Q_x is a positive definite quadratic form depending smoothly on x. Thus P_0 has the kernel representation

(15.35) $$P_0(f)(x) = \int_{\mathbb{R}^m} K_0(x, x-t, y) f(t)\, dt$$

with

$$K_0(x,u,y) = \frac{c\, y \det Q_x^*}{(y^2 + Q_x^*(u))^{m+1/2}}$$

where Q_x^* is the quadratic form corresponding to the inverse matrix; this is because of the well-known identity

$$\frac{1}{(2\pi)^m} \int_{\mathbb{R}^m} e^{-y|\xi'|} e^{i\langle x, \xi'\rangle} d\xi' = \frac{c\,y}{(y^2 + |x|^2)^{m+1/2}},$$

(see [33], p.61). Hence $\sup_{(x,y) \in \mathbb{R}_+^{m+1}} \int K_0(x,u,y)\, du \leq A$ which proves our assertion about P_0. P_1 has symbol of order -1, and so the assertion concerning it also holds.

Now let X be an allowable vector field. In an appropriate coordinate neighborhood we can write

$$X = \sum_j a_j(x,y) \frac{\partial}{\partial x_j} + b(x,y) \frac{\partial}{\partial y}$$

with $b(x, 0) = 0$, since X is tangential at the boundary, $(y=0)$. Thus if we also denote by X its restriction to the boundary, we have because of (15.6) and (15.7),

(15.36) $$X P_0 = P_0 X + P'$$

and

$$X^2 P_0 = P_0 X^2 + P'X + P'' \tag{15.37}$$

where P' and P'' are Poisson operators of order 0.

Next let $\Psi_0, \Psi_1, \ldots, \Psi_k, \ldots$ be the functions defined on $\mathbb{R}^1$, by $\hat{\Psi}_0(\xi) = \psi(\xi)$, $\hat{\Psi}_k(\xi) = \psi(\xi 2^{-k}) - \psi(\xi 2^{-k+1})$, where ψ is defined in the proof of Lemma 13.5. Then we have (as is easily verified), that Ψ_k is an even function in t, $1 \equiv \sum_{k=0}^{\infty} \hat{\Psi}_k(\xi)$, and

$$\int_{-\infty}^{\infty} |t|^{\alpha} \, |\Psi_k^{(j)}(t)| \, dt \leq A\, 2^{(j-\alpha)k}, \quad j=0,1,2, \tag{15.38}$$

where $\Psi_k^{(j)}(t) = (\frac{d}{dt})^j \, \Psi_k(t)$. Also $\int_{-\infty}^{\infty} \Psi_k(t)\, dt = 0$, $k=1,2,\ldots$.

Now let $f \in \Gamma_{\alpha}(bM)$, $0 < \alpha \leq 1$, as in [9], p. 492. We claim that $t \longrightarrow f(\gamma(t))$ is in Λ_{α} as a function of t, where γ is an integral curve of the vector field X. This can be seen by the discussion in [9], pp. 492, 493, if we define the "metric" ρ in terms of a normal coordinate system which has X as one of its basis elements. Next let $\varphi_t(x)$ denote the (local) homeomorphisms given by $\varphi_t(x) = \exp tX$, and set

$$f_k(x) = \int_{-\infty}^{\infty} f(\varphi_t(x))\, \Psi_k(t)\, dt \tag{15.39}$$

We may assume (upon multiplication by a suitable cut-off function) that f is supported in a suitably small coordinate neighborhood so that (15.39) is well defined. However,

$$\|f_k\|_{L^{\infty}} \leq \int_{-\infty}^{\infty} \|f(\varphi_t(\cdot)) - f(\cdot)\|_{L^{\infty}} \, |\Psi_k(t)| \, dt \leq A\, 2^{-\alpha k},$$

since $\|f(\varphi_t(\cdot)) - f(\cdot)\|_{L^\infty} \leq At^\alpha$ and f is in Λ_α along integral curves of X. That is we have

(15.40) $$\|f_k\|_{L^\infty} \leq A\,2^{-\alpha k}, \qquad k=0,1,2,\ldots.$$

Now $(Xf\varphi_t(x)) = \frac{d}{dt} f(\varphi_t(x))$, and so

(15.39′) $$Xf_k = -\int_{-\infty}^{\infty} f(\varphi_t(x))\,\Psi_k^{(1)}(t)\,dt,$$ and again by (15.38)

(15.41) $$\|Xf_k\|_{L^\infty} \leq A\,2^{(1-\alpha)k}.$$

Similarly,

(15.42) $$\|X^2 f_k\|_{L^\infty} \leq A\,2^{(2-\alpha)k}.$$

We have, however, that $f = \sum_{k=0}^{\infty} f_k$, since $\sum_{k=0}^{\infty} \Psi_k(t) = \delta$, because $\sum \hat{\Psi}_k(\xi) \equiv 1$. Therefore if $F = P_0(f)$, $F = \sum F_k$, with $F_k = P_0(f_k)$.

Hence the boundedness of P_0 on $L^\infty(\mathbb{R}^m)$ to $L^\infty(\mathbb{R}^{m+1})$ shows that (by (15.40))

(15.43) $$\|F_k\|_{L^\infty} \leq A\,2^{-\alpha k}.$$

Since f is in $\Gamma_\alpha(bM)$, then $f \in \Lambda_{\alpha/2}(bM)$; see [9], Theorem 20.1. Thus because of its definition (15.39)

(15.44) $$\|f_k\|_{\Lambda_{\alpha/2}} \leq A.$$

Again because $f \in \Gamma_\alpha(bM)$, (and $f \in \Lambda_{\alpha/2}(bM)$), we have

$$|\{f(\varphi_t(x)) - f(x)\} - \{f(\varphi_t(y)) - f(y)\}| \leq A \min \{|t|^\alpha, |x-y|^{\alpha/2}\}$$

$\leq A t^{\alpha/2} |x-y|^{\alpha/4}$. Therefore by (15.39′)

$$\|X f_k\|_{\Lambda_{\alpha/4}} \leq A\, 2^{(1-\alpha/2)k}. \tag{15.45}$$

By (15.36),

$$XF_k = P_0 X f_k + P' f_k .$$

The L^∞ boundedness of P_0 gives $\|P_0 X f_k\|_{L^\infty} \leq A\, 2^{(1-\alpha)k}$.

By Lemma 13.12, $\|P' f_k\|_{L^\infty} \leq \|P' f_k\|_{\Lambda_{\alpha/4}} \leq A \|f_k\|_{\Lambda_{\alpha/4}} \leq A$ (see (15.44)). Therefore,

$$\|X F_k\|_{L^\infty} \leq A\, 2^{(1-\alpha)k}, \quad \text{if } 0 < \alpha \leq 1. \tag{15.46}$$

$X^2 F_k$ is handled the same way, using (15.37) and (15.45). The result is

$$\|X^2 F_k\|_{L^\infty} \leq A\, 2^{(2-\alpha)k}, \qquad 0 < \alpha \leq 1. \tag{15.47}$$

The combination of (15.43), (15.45), and (15.47) shows that $\sum F_k = F = P_0(f)$ belongs to Λ_α on any integral curve of X, by Lemma 13.1 (and (13.4)).

To summarize: If $f \in \Gamma_\alpha(bM)$, $0 < \alpha \leq 1$, then $P_0(f) \in \Lambda_\alpha(\overline{M})$. However $P = P_0 + P_1$ and P_1 is a Poisson operator of order -1; thus $\frac{\partial}{\partial x_j} P_1$ and $\frac{\partial}{\partial y} P_1$ are operators of order 0, and since $f \in \Lambda_{\alpha/2}(bM)$ if $f \in \Gamma_\alpha(bM)$, we have that $P_1(f) \in \Lambda_{1+\alpha/2}(\overline{M}) \subset \Gamma_\alpha(\overline{M})$, if $\alpha \leq 1$. Therefore we have proved that if $f \in \Gamma_\alpha(bM)$ then $P(f) \in \Gamma_\alpha(\overline{M})$, when $0 < \alpha \leq 1$.

The rest of Lemma 13.4 is then proved by induction, using the recursive definition of $\Gamma_\alpha(\overline{M})$ and the identity (15.36) where X is any allowable vector field.

We prove next that the approximate Neumann operator N_a maps Λ_α to $\Gamma_{\alpha+2}$. (We already know that it maps Λ_α to $\Lambda_{\alpha+1}$.) Let us do this first in the case $n \geq 2$. By the form of N_a given in (15.3), and since $\Lambda_{\alpha+2} \subset \Gamma_{\alpha+2}$, it suffices to consider the term $P\,S_2 T_1 \,\text{Rest}\, D_1 G$. Now $T_1 \,\text{Rest}\, D_1 G$ maps Λ_α to Λ_α, and S_2 maps this to $\Lambda_{\alpha+1}$ (see Theorem 14.3 (b)). Moreover $\Lambda_\alpha \subset \Gamma_\alpha$ and S_2 maps Γ_α to $\Gamma_{\alpha+2}$. (For this, see [9], p. 465 and p. 492.) Thus an application of Lemma (15.34) shows that N_a maps Λ_α to $\Gamma_{\alpha+2}$.

The argument for $n=1$ requires that instead of using the approximate Neumann operator given by (10.29) and (10.23) (which has an extraneous $T_0 = E^+$ term in it), we use the form given by (10.32). There the $\overline{K}$ operator is of type S_2, and the argument for the term that involves it goes as before. The "error term" is then of the form

$$P(\overline{C}_b T_{-1} \,\text{Rest}\, T_1 G(f)).$$

If $f \in \Lambda_\alpha$, then $T_{-1} \,\text{Rest}\, T_1 G(f) \in \Lambda_{\alpha+2} \subset \Gamma_{\alpha+2}$. However $\overline{C}_b$ is an operator of type 0, and so maps $\Gamma_{\alpha+2}$ to itself, by [9], pp. 465 and 492. So the result for N_a is proved in this case also.

We already know (by Theorem 15.30) that $u = N(f) \in \Lambda_{\alpha+1}$. Hence our previous arguments show that $R(u) \in \Gamma_{\alpha+2}$, and the proof of Theorem 15.33 is complete.

Chapter 16. Solution of $\bar{\partial} U = f$

We begin by pointing out that our solution of the $\bar{\partial}$-Neumann problem is, strictly in the interior of M, elliptic in the sense that there is a gain of two in the usual sense. This of course follows by the general "interior regularity" of solutions of elliptic equations, but in our case it is a consequence of Theorems 15.1, part (b), and Theorem 15.33, since allowable vector fields are not restricted away from the boundary.

Another fact is that the "normal" component of the solution behaves in an elliptic way even up to the boundary. This can be made precise as follows. Let ω_{n+1} be a smooth (1,0) form which near the boundary is given by $\omega_{n+1} = \sqrt{2}\,\partial\rho$. (What it is away from the boundary is irrelevant.) In terms of it we have the decomposition, near the boundary, for any (0,1) form u,

$$u = u_t + \nu(u)\,\bar{\omega}_{n+1}$$

where $\nu(u)$ is the component of $\bar{\omega}_{n+1}$, and u_t is orthogonal to $\bar{\omega}_{n+1}$. In general we put $\nu(u) = \bar{\omega}_{n+1} \lrcorner\, u$. The Dirichlet boundary condition $\nu(u)|_{bM} = 0$ of (11.2) is responsible for the following regularity property.

<u>16.1 Theorem</u>. <u>The mapping</u> $f \longrightarrow \nu(N(f))$ <u>maps</u> $L^p_k(\bar{M})$ <u>to</u> $L^p_{k+2}(\bar{M})$ <u>and</u> $\Lambda_\alpha(\bar{M})$ <u>to</u> $\Lambda_{\alpha+2}(\bar{M})$, <u>if</u> $\alpha > 0$. <u>It also maps</u> $L^\infty(\bar{M})$ <u>to</u> $\Lambda_2(\bar{M})$.

<u>Proof</u>. It is sufficient to prove that $\nu(N_a)$ has the same boundedness properties as those claimed for $\nu(N)$ and that $\nu(R)$ maps L^p_{k+1} to L^p_{k+2}, and $\Lambda_{\alpha+1}$ to $\Lambda_{\alpha+2}$ (since we already know that $u \in L^p_{k+1}(\bar{M})$, or

$u \in \Lambda_{\alpha+1}(\overline{M})$, under our hypotheses). We shall need the following observation

16.2 Lemma. On the subspace of g, such that $\nu(g) = 0$, $\nu(P)$ is a Poisson operator of order -1.

In fact the main term of P is a diagonal operator. The non-diagonal part of P has symbol of order -1 (see (7.55)).

Thus the arguments we have already presented in detail above show that $\nu(N_a)$ and $\nu(R)$ have the required regularity properties and the theorem is proved.

We come now to the problem

$$\bar{\partial} U = f \tag{16.3}$$

where f is a given $(0,1)$ form which satisfies

$$\bar{\partial} f = 0 \tag{16.4}$$

in the weak sense, i.e., $((\bar{\partial})^* \varphi, f) = 0$, for any $\varphi \in C^\infty_{(0,1)}(\overline{M})$, which has compact support in M.

Whenever $f \in L^2(\overline{M})$, f satisfies (16.4) and f is orthogonal to the Harmonic forms, i.e., $(f, \psi) = 0$, $\psi \in \mathcal{H}$, then by the formalism of the $\bar{\partial}$-problem (see e.g. [8], p. 52), we know that

$$U = \vartheta N(f) \tag{16.5}$$

is the unique weak solution of (16.3) with the property that

$$(U, F) = 0 \tag{16.6}$$

for all holomorphic function F which are in $L^2(\overline{M})$.

Observe that by the regularity theorems already proved for N,

$U = \vartheta N(f) \in L^p_k(\overline{M})$, if $f \in L^p_k(\overline{M})$, and $U \in \Lambda_\alpha(\overline{M})$ if $f \in \Lambda_\alpha(\overline{M})$.

16.7 Theorem. Suppose f is a $(0,1)$ form with $\bar\partial f = 0$, and $f \perp \mathcal{H}$. Then U, given by (16.5), is the unique (weak) solution of $\bar\partial U = f$ which satisfies (16.6), for all holomorphic functions $F \in L^{p'}(\overline{M})$, when $f \in L^p_k(\overline{M})$, with $1/p + 1/p' = 1$ If $f \in \Lambda_\alpha(\overline{M})$, or $f \in L^\infty(\overline{M})$, then the solution U of (16.3) is determined uniquely by $(U, F) = 0$ for all holomorphic F in $L^2(\overline{M})$. Moreover

(a) U, and $XU \in L^p_k(\overline{M})$, if $f \in L^p_k(\overline{M})$, where X is any allowable vector field.

(b) $U \in L^p_{k+1/2}(\overline{M})$, if $f \in L^p_k(\overline{M})$

(c) $U \in \Gamma_{\alpha+1}(\overline{M}) \cap \Lambda_{\alpha+1/2}(\overline{M})$ if $f \in \Lambda_\alpha(\overline{M})$ and $\alpha > 0$. If $f \in L^\infty(\overline{M})$ then $U \in \Lambda_{1/2}(\overline{M})$.*

Proof. We shall restrict our attention to $\overline{M} \cap U$, where U is a sufficiently small open set in M', as we have done systematically above and make our estimates for φu, φU, etc., where $\varphi \in C^\infty$ and has compact support in U. Thus $U \in L^p_{k+1/2}(\overline{M})$ will mean that $\varphi U \in L^p_{k+1/2}(\mathbb{R}^{2n+2}_+)$, when by an appropriate coordinate system we have identified $\overline{M} \cap U$ with a neighborhood of the origin in $\overline{\mathbb{R}}^{2n+2}_+$.

In U we choose an orthonormal frame $\omega_1, \omega_2, \ldots, \omega_{n+1} = \sqrt{2}\,\partial\rho$, and its dual frame $Z_1, Z_2, \ldots, Z_{n+1}$. Then we know (see Chapter 4) that

$$\vartheta u = -\sum_{j=1}^{n+1} Z_j(u_j) + \sum_{j=1}^{n+1} h_j u_j \tag{16.8}$$

* See also the discussion in Chapter 17.

where $u = \sum_{j=1}^{n+1} u_j \bar{\omega}_j$.

Notice that $\nu(u) = u_{n+1}$, and so part (a) follows from Theorem (16.1), and Theorem 15.1, part (b), since the Z_j are allowable vector fields if $1 \leq j \leq n$.

Now if X is any allowable vector field, we already know (by Theorem 15.1, parts (a) and (b)), that $u_j \in L^p_{k+1}$ and $X^2 u_j \in L^p_k$, locally. We can introduce new coordinates, so that $X = \frac{\partial}{\partial x_1}$, and after multiplication with suitable cut-off functions we have

(16.9) $$u_j \in L^p_{k+1}(\mathbb{R}^{2n+2}_+) \quad \text{and} \quad \frac{\partial^2}{\partial x_1^2} u_j \in L^p_k(\mathbb{R}^{2n+2}_+)$$

We claim that as a consequence of (16.9) we have

$$\frac{\partial u}{\partial x_1} \in L^p_{k+1/2}(\mathbb{R}^{2n+2}_+)$$

In fact, let F be equal to the extension of u_j to all of $\mathbb{R}^{2n+2}$, as in (7.58). Then because of (16.9), and the commutativity of $\sim$ with $\frac{\partial}{\partial x_1}$, we get

(16.10) $$F \in L^p_{k+1}(\mathbb{R}^{2n+2}) \quad \text{and} \quad \frac{\partial^2 F}{\partial x_1^2} \in L^p_k(\mathbb{R}^{2n+2}).$$

We claim that as a result

(16.11) $$\frac{\partial F}{\partial x_1} \in L^p_{k+1/2}(\mathbb{R}^{2n+2}),$$

Now let $J^{(1)}_z$ be the operator which is given by multiplication on the Fourier transform side by the function $(1 + |\xi_1|^2)^{-z/2}$.

We know (see [4]) that J_{iy} is a bounded operator on $L^p(\mathbb{R}_1)$ (and hence on $L^p(\mathbb{R}^{2n+2})$) for all real y with polynomial growth in y. Thus if $F \in L^p_{k+1}$ then so is $F_{iy} = J_{iy} F \in L^p_{k+1}$, with norm growing at most polynomially in y. Next $F_{-2+iy} = J_{iy}(1 - (\frac{\partial}{\partial x_1})^2) F \in L^p_k$, with norm again polynomially growing at worst in y, since $(\frac{\partial}{\partial x_1})^2 F \in L^p_k$. Thus by the convexity argument in [4], we have $F_{-1} = J_{-1}(F) \in L^p_{k+1/2}$. However, $\frac{\partial}{\partial x_1} = A + BJ_{-1}$, where the operators are bounded on $L^p(\mathbb{R}^1)$ (and hence on $L^p_{k+1/2}$, or L^p_{k+1}), by the Marcinkiewicz multiplier theorem. (A can be taken to be multiplication on the Fourier transform side by a smooth function of compact support. B corresponds to a multiplier which is smooth and which for large ξ_1 equals $\frac{i\xi_1}{(1+|\xi_1|^2)^{1/2}}$.) Thus $F \in L^p_{k+1/2}(\mathbb{R}^{2n+2})$. Going back to the definitions we see that each Xu_j belongs locally to $L^p_{k+1/2}$, and so part (b) of the theorem is also proved.

The fact that $U \in \Gamma_{\alpha+1}(\overline{M})$ if $f \in \Lambda_\alpha(\overline{M})$ follows directly from Theorem 15.33. The fact that $U \in \Lambda_{\alpha+1/2}(\overline{M})$ is already implicit in what we have previously done. In fact the P term in N_a (see (15.3)) involves an operator of type S_2. If we apply an allowable vector field to this P term we get a similar term, but with S_2 replaced by S_1. We then only need use Theorem 14.3, part (b) to complete the argument. The details are so similar to previous arguments that they may be left to the reader.

Chapter 17. Concluding Remarks

In this chapter we point out some further results in order to round out the picture we have presented above. We give only an indication of the proofs, since the reader who has followed us this far should have no difficulty in filling out the required details.

a. The domain of $\square$

The first question we pose is that of giving a characterization of those u, in terms of regularity conditions on u and boundary conditions, so that u belongs to the self-adjoint extension $\square_e$ of $\square$ (described in Chapter 11) i.e., when is $u = N(f)$, $f \in L^2(\overline{M})$; or more generally when is $u = N(f)$, where $f \in L^p_k(\overline{M})$?

Observe first that by Theorem 15.1, if $f \in L^p_k(\overline{M})$, then

$$(17.1) \qquad p(Z, Z)u \in L^p_k(\overline{M})$$

where p is any polynomial of second degree in the allowable vector fields

Also

$$(17.2) \qquad \overline{Z}_{n+1}\, u \in L^p_{k+1}(\overline{M})$$

where Z_{n+1} is a holomorphic vector field which near the boundary equals $\frac{1}{\sqrt{2}}(\frac{\partial}{\partial \rho} + iT)$.

Incidentally, conclusion (a) of Theorem 15.1, namely that

$$(17.3) \qquad u \in L^p_{k+1}(\overline{M}),$$

is a consequence of (17.1) and (17.2).

In fact, let $Z_1, Z_2, \ldots, Z_{n+1}$ be a basis of the holomorphic vector fields (in an appropriate coordinate neighborhood of the boundary). Then the derivatives of u along the real and imaginary parts of $Z_1, \ldots, Z_n$ of u belong to $L^p_k(\overline{M})$ because, Z_j are allowable vector fields, when $1 \leq j \leq n$. Moreover by the commutation relation (Lemma 6.20), Tu again belongs to $L^p_k(\overline{M})$. Finally because of (17.2) $\frac{\partial}{\partial \rho} u \in L^p_k(\overline{M})$, and so $u \in L^p_{k+1}(\overline{M})$. Observe also that if (17.1) and (17.2) are satisfied (then because of (17.3)) and the explicit form of $\square$ given in (6.23), we have

(17.4) $$\square u \in L^p_k(\overline{M}).$$

Finally since $u \in L^p_{k+1}(\overline{M}) \subset L^p_1(\overline{M})$, it follows that its restriction to the boundary is a well-defined element in $B^p(bM)$, (see Theorem 14.1)). Thus $\nu(u)|_{bM}$ is well-defined and it can be proved by a simple limiting argument that

(17.5) $$\nu(u)|_{bM} = 0.$$

Similarly since $\overline{Z}_{n+1} u \in L^p_{k+1}(\overline{M}) \subset L^p_1(\overline{M})$ then in view of (6.16), $\nu(\bar{\partial} u)|_{bM} \in B^p(bM)$, and again

(17.6) $$\nu(\bar{\partial} u)|_{bM} = 0.$$

The considerations above can be summarized as follows.

<u>17.7 Theorem.</u> <u>Suppose</u> u <u>is given on</u> M. <u>Then</u> $u = N(f)$, <u>where</u> $f \in L^p_k(\overline{M})$, <u>if and only if</u> u <u>satisfies the regularity conditions</u> (17.1) <u>and</u> (17.2), <u>together with the boundary conditions</u> (17.5) <u>and</u> (17.6).

b. Solution of $\bar{\partial}U = f$ when f is bounded

According to Theorem (16.7), the mapping $f \longrightarrow \vartheta N(f) = U$ (which solves $\bar{\partial}U = f$ whenever this is possible) is bounded from $L^\infty(\overline{M})$ to $\Lambda_{1/2}(\overline{M})$. In keeping with the results we have proved for $\alpha > 0$ we wish to also show that it is bounded from $L^\infty(\overline{M})$ to $\Gamma_1(\overline{M})$. In order to state the result it is convenient to give the space $\Gamma_1(\overline{M})$ a norm.* To do this we first give the allowable vector fields on $\overline{M}$ a norm. We consider a finite covering of $\overline{M}$ by coordinate patches, and in each such patch we express $X = \sum a_j(x) \frac{\partial}{\partial x_j}$. Then the norm of X, $\|X\|$, is the supremum of the derivatives of order not greater than one of the $a_j(x)$, taken over the various coordinate patches.

Finally the Γ_1 norm of u is the supremum of the Λ_1 norms of all functions $t \longrightarrow u(\gamma(t))$, where γ ranges over all segments of integral curves of allowable vector fields of norm ≤ 1.

17.8 Theorem. The mapping $f \longrightarrow U = \vartheta N(f)$, defined for those $f \in C^\infty(\overline{M})$, so that $\bar{\partial}f = 0$, and $f \perp \mathcal{K}$, satisfies

$$\|U\|_{\Gamma_1 \cap \Lambda_{1/2}} \leq A\|f\|_{L^\infty}. \tag{17.9}$$

We will now review the background of this theorem.

In the case when $\overline{M}$ is a sub-domain of $\mathbb{C}^{n+1}$, Henkin [13] constructed an integral operator $f \longrightarrow H(f)$, so that $\bar{\partial}H(f) = f$, when $\bar{\partial}f = 0$, and for which

*In our previous discussion we did not give the Γ_α spaces norms. However, if we define their norm (similarly to the Γ_1 norm described below) then Theorem 15.33, and Theorem 16.7 (c), can be restated in terms of boundedness in these norms.

Henkin and Romanov [14] proved the estimate

(17.10) $$\|H(f)\|_{\Lambda_{1/2}} \leq A \|f\|_{L^\infty} .$$

The latter followed earlier results of Grauert and Lieb [10], Kerzman [20], and others about similar operators. A little later one of us introduced the Γ_α spaces in [34]*; it was also asserted that an estimate like (17.9), but slightly weaker, held for $H(f)$. Actually the proof we had in mind only showed something even weaker (namely that $H(f)$ is in $\Lambda_{1-\varepsilon}$ in the "good directions"). The details of that proof, together with other "sharp" results for $H(f)$, appear in the thesis of S. Krantz [23].

To prove Theorem 17.8 we shall use still another solution of the problem $\bar\partial U = f$, the one which is studied by Phong in his forthcoming dissertation [28]. This solution, $f \longrightarrow \Phi(f)$, is the one characterized by $\bar\partial\Phi(f) = f$ with $\Phi(f)$ orthogonal to holomorphic functions, the orthogonality being in terms of integration taken on the boundary. Following a suggestion of Kohn, one can give a simple expression for $\Phi(f)$ in terms of solutions of the corresponding $\bar\partial_b$ problem

In fact we shall see that the estimate (17.9) holds not only for the solution $\vartheta N(f)$, but also for $H(f)$, and $\Phi(f)$. The fact that it holds for $H(f)$ was also proved recently by Henkin,** but by different methods.

We first sketch the proof of Theorem (17.9) when $n \geq 2$. Let C

*The definitions given in [34] for Γ_α have undergone a notational change. The Γ_α spaces used here correspond to the $\Gamma_{\alpha/2}$ spaces in [34]. Also our spaces Γ_α are called $\Gamma_{\alpha/2,\alpha}$ in Krantz [23].

**Personal communication.

denote the Cauchy-Szegö projection operator, which by the formalism for the boundary complex (see e.g. [8], Chapter V) can be written

(17.11) $$C(v)|_{bM} = v - \vartheta_b G_b \bar{\partial}_b v$$

where G_b is the "Neumann operator" for the $\bar{\partial}_b$ complex. From the regularity properties of G_b it follows that if $v \in C^\infty(bM)$ then $C(v)|_{bM} \in C^\infty(bM)$, and hence the holomorphic function $C(v) \in C^\infty(\overline{M})$.

Now set

$$\Phi(f) = \vartheta N(f) - C(U_b)$$

where $U = \vartheta N(f)$, and $U_b = U|_{bM}$.

Then the function $w = \Phi(f)$ satisfies $\bar{\partial} w = f$ and w is orthogonal on bM to holomorphic functions. From this it follows that

(17.12) $$\square^{(0)} w = \vartheta \bar{\partial} w = \vartheta f$$

and if $w_b = w|_{bM}$, then

(17.13) $$\bar{\partial}_b w_b = f_b$$

where f_b is the "restriction" of the (0.1) from f to bM. Now $\|f_b\|_{L^\infty(bM)} \leq \|f\|_{L^\infty(M)}$, and since w_b is the solution of (17.13) orthogonal to holomorphic functions, it follows from [9], §17 that

(17.14) $$\|w_b\|_{\Gamma_1} \leq A\|f\|_{L^\infty}.$$

Similarly since (except for smoother terms) w_b is expressed in terms of f_b by operators of type III_1, we also have by Theorem 14.3, (b)

(17.15) $$\|w_b\|_{\Lambda_{1/2}} \leq A\|f\|_{L^\infty}.$$

Now by an analogue of what was done in Chapter 7 for the Dirichlet problem for $\square^{(0)}$ we have

$$w = \widetilde{G}(\square^{(0)} w) + \widetilde{P}(w_b)$$

where $\widetilde{G}$ and $\widetilde{P}$ are appropriate Green and Poisson operators.

Since $\square^{(0)} w = \vartheta f$, we get by elliptic estimates (of the type we have already made) that

$$(17.16) \qquad \|\widetilde{G}(\square^{(0)} w)\|_{\Gamma_1} \leq C\|\widetilde{G}(\square^{0}(w))\|_{\Lambda_1} \leq C\|f\|_{L^\infty}.$$

Next analogue of Lemma (15.34) together with (17.14) and (17.15) shows that

$$\|\widetilde{P}(w_b)\|_{\Gamma_1 \cap \Lambda_{1/2}} \leq A\|f\|_\infty .$$

Putting these together gives

$$(17.17) \qquad \|\Phi(f)\|_{\Gamma_1 \cap \Lambda_{1/2}} \leq A\|f\|_\infty .$$

Our theorem will then be proved if we can show that the holomorphic function $\vartheta N(f) - \Phi(f)$, satisfies

$$(17.18) \qquad \|\vartheta N(f) - \Phi(f)\|_{\Gamma_1} \leq A\|f\|_\infty .$$

This is a consequence of the following general fact:

17.19 Lemma. Suppose F is holomorphic in M and belongs to $\Lambda_{1/2}(\overline{M})$. Then $F \in \Gamma_1(\overline{M})$, and $\|F\|_{\Gamma_1} \leq A\|F\|_{\Lambda_{1/2}}$.

This result is actually true in much more general setting. First, no pseudo-convexity hypotheses on the boundary need be made. Secondly, there is an analogue where $\Lambda_{1/2}$ and Γ_1 are replaced by Λ_α and $\Gamma_{2\alpha}$

respectively for $0 < \alpha < \infty$. That was announced by one of us in [34]; since details have not yet appeared we shall give the proof in the case that is needed, namely $\alpha = 1/2$. The general case can be proved similarly

Everything is based on the following two simple observations. Suppose P is a point in M and the distance of P from bM is δ. Then any complex one-dimensional disc, centered at P of radius δ lies in M. However, if the one-dimensional disc lies along a "good direction" then even if it had a much longer radius it would still lie in M. To be precise, there exists positive constants c_1 and c_2, so that if X is any allowable vector field (of norm ≤ 1, as defined above), then

(17.20) The one-disc, centered at P, in the direction $\mathbb{C} X_P$, and of radius $c_1 \delta^{1/2}$, lies in M.

Moreover this disc is at a distance at least $c_2 \delta$ from bM.

(17.20) is a direct consequence of the definition of an allowable vector field, since $X|_{bM} \in T^{1,0}_{(bM)} \oplus T^{0,1}_{(bM)}$.

The second observation is an easy consequence of Cauchy's integral formula. We suppose that f is a holomorphic function of one complex variable defined in the disc D_r of radius r, centered at the origin.

(17.21) If $|f| \leq 1$ in D_r, then $|f'(0)| \leq c/r$.

(17.21′) If $\|f\|_{\Lambda_{1/2}(D_r)} \leq 1$, then $|f'(0)| \leq c/r^{1/2}$.

We combine these two observations as follows. First

(17.22) $|(YF)(z)| \leq A\, \delta^{-1/2}$

if δ denotes the distance of z from bM, where Y is any vector field. Similarly, if Y_1 and Y_2 are any two vector fields $|Y_1 Y_2 F(z)| \leq A\delta^{-3/2}$. These follow by applying (17.21) and (17.21′) with discs of radius δ. Next if X is any allowable vector field (of norm ≤ 1), then using discs of radius $c_1\delta^{1/2}$ we get from the above

(17.23) $$|YX^2F| \leq A\delta^{-3/2}.$$

Let Y be the vector field $\frac{\partial}{\partial\rho}$; since $\rho = \delta$ an integration in ρ in (17.23) yields

(17.24) $$|X^2F| \leq A\delta^{-1/2}.$$

Let $\gamma(t)$ be an integral curve of the vector field X. We must show that F restricted to $\gamma(t)$ is in Λ_1, i.e.,

(17.25) $$(F(\gamma(t)) + F(\gamma(-t)) - 2F(\gamma(0))| \leq A|t| \quad \text{for } |t| \leq t_0.$$

For any $h > 0$, we consider another integral curve $\gamma_h(t)$ of the vector field X, which is determined by the fact that $\gamma_h(0)$ is the point obtained by moving a distance h along a geodesic normal to bM which passes through $\gamma(0)$; i.e., the distance of $\gamma_h(0)$ from bM is the distance of $\gamma(0)$ from bM plus h, and the distance of $\gamma_h(0)$ from $\gamma(0)$ is h. Now it is easy to see that

(17.26) $$|\gamma_h(t) - \gamma(t)| \leq Ah, \quad \text{if } |t| \leq t_0,$$

and since X is tangential at bM, then

(17.27) $$\text{distance } \gamma_h(t) \text{ from bM} \geq ch \quad \text{if } |t| \leq h^{1/2}.$$

By (17.22) and (17.26) it then follows that

(17.28) $$|F(\gamma(t)) - F(\gamma_h(t))| \leq Ah^{1/2}, \quad |t| \leq t_0 \quad \text{if} \quad |t| \leq h^{1/2};$$

the same inequality also holds if t is replaced by $-t$, or 0. Finally,

$$|F(\gamma_h(t)) + F(\gamma_h(-t)) - 2F(\gamma_0(0))| \leq t^2 \sup|X^2F|$$

where the sup is taken over the curve γ_h, between $-t$ and t. Using (17.24) and (17.27) we have

$$|F(\gamma_h(t)) + F(\gamma_h(-t)) - 2F(\gamma_h(0))| \leq At^2 h^{-1/2}$$

Finally we take $h^{1/2} = |t|$, and when this is combined with (17.28) we obtain (17.25) which is our desired conclusion. If we keep track of the constants we see that Lemma 17.19 is proved.

The proof of Theorem 17.8 when $n \geq 2$ is then concluded by applying the lemma to $F = \Phi(f) - \vartheta N(f)$.

The proof in the case $n=1$ requires an additional argument; but here we shall be very sketchy. The identity (3.19) gives us an operator K_1 which is convolution on the Heisenberg group H_1 by a kernel homogeneous of degree -3 (so that K_1 is of type 1) and so that

(17.29) $$K_1 \bar{\partial}_b = I - C_b$$

where C_b is the Cauchy-Szegö projection. The operator K_1 can now be "induced" to bM (see the discussion in the "Guide to Part II," and in Chapter 10). The result is an operator $\widetilde{K}_1$ of type 1, so that now

(17.30) $$\widetilde{K}_1 \bar{\partial}_b = I - \widetilde{C}_b + S_1 + S.$$

$\widetilde{C}_b$ is the Cauchy-Szegö kernel for bM. This identity can be verified as a consequence of (17.29), if we use Fefferman's asymptotic description of the Cauchy-Szegö (see [7], and also [3]). The error term S_1 is the usual kind of operator of type 1. S is a further error term involving a logarithmic singularity in the kernel (and hence is in fact better behaved than S_1). This logarithmic singularity is due to the presence of a logarithmic term in the description of the Cauchy-Szegö kernel. Suppose now (17.30) is verified.

If $\bar{\partial}_b w_b = f_b$, with w_b orthogonal to holomorphic functions, we get from (17.30)

(17.31) $$\widetilde{K}_1 f_b = w_b + (S_1 + S) w_b \ .$$

However, if the L^∞ norm of f is controlled, then we know that the $\Lambda_{1/2}$ norm of $U = \vartheta N(f)$ is controlled, and thus so is the $L^2(bM)$ norm of $U_b - C_b(U_b) = w_b$. Since $\widetilde{K}_1$ is of type 1 (and hence is bounded from $L^\infty(bM)$ to $\Lambda_{1/2}(bM)$ and $\Gamma_1(bM)$) we obtain after iterating (17.31) several times that

$$\|w_b\|_{\Gamma_1} \leq A\|f\|_{L^\infty}, \text{ and } \|w_b\|_{\Lambda_{1/2}} \leq A\|f\|_\infty .$$

The proof of Theorem 17.8 for the case n=1 is then completed as the case $n \geq 1$ discussed above.

Other regularity estimates for solutions of $\bar{\partial}U = f$ which have some bearing on our results in Theorem 16.7(c), and Theorem **17.8** are contained Siu [32], Krantz [32], Alt [1], and Øverlid [27].

c. Some examples

We shall show by some simple examples that our isotropic results concerning regularity for the Λ_α spaces can not be improved.

We consider first the equation $\bar\partial U = f$.

Let $\overline{M}$ denote the unit ball in $\mathbb{C}^2$. Then it is easy to verify that the usual (Euclidean) metric on $\mathbb{C}^2$ is a Levi metric for $\overline{M}$. Let $U(z_1, z_2) = \bar z_1 f_0(z_2)$, where f_0 is holomorphic in z_2 $(|z_2| < 1)$. Then $\frac{\partial U}{\partial \bar z_1} = f_0(z_2)$ and $\frac{\partial U}{\partial \bar z_2} = 0$, so $\bar\partial U = f$, with $f = f_0\, d\bar z_1$. Observe also that U is orthogonal to holomorphic functions in $\overline{M}$. Now if $f_0(z_2) = (1-z_2)^{i\gamma}$, with γ real, $\gamma \neq 0$, then f is bounded, but $U \notin \Lambda_{1/2+\epsilon}$, $\epsilon > 0$. (Merely test the Lipschitz character of U at the point $(0,1)$, and use the fact that $|z_2| \approx (1-|z_1|)^{1/2}$ as $(z_1, z_2) \longrightarrow (0,1)$ in $\overline{M}$.) Similarly if $f_0(z_1) = (1-z_2)^{\alpha+i\gamma}$, then $f \in \Lambda_\alpha$, but $U \notin \Lambda_{\alpha+1/2+\epsilon}$, $\epsilon > 0$.*

For the $\bar\partial$-Neumann problem it is convenient to deal with the domain $\mathcal{D} = \{\operatorname{Im} z_2 > |z_1|^2\}$.

Let $H_1 = \{(\zeta, t),\ \zeta \in \mathbb{C}^1,\ t \in \mathbb{R}^1\}$ denote the Heisenberg group. It acts on $\mathcal{D}$ as holomorphic self-mappings by

$$(z_1, z_2) \longrightarrow (z_1 + \zeta,\ z_2 + t + 2i\bar\zeta z_1 + i|\zeta|^2).$$

We shall construct a Levi-metric on $\mathcal{D}$ which is invariant under the action of H_1, and for which $\rho = \operatorname{Im} z_2 - |z_1|^2$ gives the geodesic distance from the boundary.

---

*Variants of this example have been considered by Kerzman [20] and Krantz [23].

In fact take $\omega^1 = dz_1$, $\omega^2 = \sqrt{2}\,\partial\rho = \sqrt{2}\,\bar{z}_1\,dz_1 - \frac{i}{\sqrt{2}}\,dz_2$. Then the ω^j are invariant under H_1, and thus so are their duals,
$Z_1 = \frac{\partial}{\partial z_1} + 2i\bar{z}_1 \frac{\partial}{\partial z_2}$, $Z_2 = i\sqrt{2}\frac{\partial}{\partial z_2}$.

We define the metric by requiring that ω^1 and ω^2 are orthonormal. Then an easy computation shows that the matrix of $\langle dz_i, dz_j\rangle$ is given by $\begin{pmatrix} 1 & -2iz_1 \\ 2i\bar{z}_1 & 2+4|z_1|^2 \end{pmatrix}$. If we take $\operatorname{Re} z_1$, $\operatorname{Im} z_1$, $t = \operatorname{Re} z_2$, and ρ as coordinates then one sees that in this metric $ds^2 = A + d\rho^2$, where A is a quadratic form in $\operatorname{Re}(dz_1)$, $\operatorname{Im}(dz_1)$, dt, with coefficients independent of ρ. From this it follows that ρ represents the geodesic distance from the boundary. Computing the determinant of the above matrix it follows that the volume element is $dz_1 \wedge d\bar{z}_1 \wedge dt \wedge d\rho$. Thus $Z_j^* = -Z_j$, $j=1,2$, and the h_j of (6.8) are zero. In addition, since $d\omega^j = 0$, $j=1,2$, the S matrix of (6.12) also vanishes. Hence if $u = \omega_1\bar{\omega}_1 + u_2\bar{\omega}_2$ then by the derivation of (6.23) yields

$$\square(u) = (-Z_1\bar{Z}_1 - Z_2\bar{Z}_2 - 2i\,(\frac{\partial}{\partial\bar{z}_2} - \frac{\partial}{\partial z_2}))\,u_1\bar{\omega}_1$$
$$+ (-Z_1\bar{Z}_1 - Z_2\bar{Z}_2)\,u_2\bar{\omega}_2\,.$$

In particular $\square$ is diagonal on (0,1) forms. The $\bar{\partial}$-Neumann boundary conditions (6.10) and (6.16) are respectively given by

$u_2|_{\rho=0} = 0$, and $\frac{\partial u_1}{\partial\bar{z}_2}\Big|_{\rho=0} = 0$. Take for example $u_1 = z_2^{\alpha+i\gamma+1}$, with γ real; $\gamma \neq 0$, and $u_2 \equiv 0$. Then u satisfies the boundary conditions and

and $\Box u = f$, where $f = f_1\bar{\omega}_1 + f_2\bar{\omega}_2$ with $f_1 = 2i(\alpha+i\gamma+1)z_2^{\alpha+i\gamma}$, $f_2 = 0$. So $f \in \Lambda_\alpha$ near the origin, while $u \notin \Lambda_{\alpha+1+\epsilon}$ there. Similarly, if $\alpha = 0$, $f \in L^\infty$ near the origin, but $u \notin \Lambda_{1+\epsilon}$ there.

The domain $\mathfrak{D}$ is unbounded, but we can modify this example as follows. Let M be a bounded domain with smooth strongly pseudo-convex boundary, so that M agrees with $\mathfrak{D}$ in some neighborhood of the origin. Construct on M a Levi-metric which agrees with the one coming from $\mathfrak{D}$ near the origin. Let $\Box'$ be the Laplacian corresponding to this metric, then $\Box'$ agrees with $\Box$ near the origin. Suppose that N is the Neumann operator for M. We let $v = N(f)$. Then $\Box'(v-u) \in C^\infty$ near the origin, and $v-u$ satisfies the boundary conditions. Hence by Proposition 9.26, and the fact that $R: \Lambda_{\alpha+1} \to \Lambda_{\alpha+3/2}$ (see the proof of Theorem 15.30) we see that $v-u \in \Lambda_{\alpha+3/2}$. Thus $v \notin \Lambda_{\alpha+1+\epsilon}$ or $\Lambda_{1+\epsilon}$, if $\alpha = 0$, and so we have our counterexample in M also.

Principal notations

Symbol	Explanation	Chapter of first occurrence
H_n	Heisenberg group	1
$\mathcal{L}_\alpha$	a left-invariant second-order operator on H_n	1
G_α $(= \Phi_\alpha)$	fundamental solution for $\mathcal{L}_\alpha$	1, 2
C_b	Cauchy-Szegö projection	3
$u(\xi, \eta)$	admissible coordinate mapping	4
$\langle , \rangle$	Levi metric	5
$\square$	$= \bar{\partial}\vartheta + \vartheta\bar{\partial}$ on (0, 1) forms	6
ρ	distance from boundary	6
$\omega_1, \omega_2, \ldots, \omega_{n+1} = \sqrt{2}\,\partial\rho$	orthonormal frame of (1, 0) forms	6
$Z_1, \ldots, Z_{n+1}$; $Z_{n+1} = \frac{1}{\sqrt{2}}\left(\frac{\partial}{\partial\rho} + iT\right)$	dual holomorphic vector fields	6
$\nu(u)$	"normal" component of form	6
T_k	standard pseudo-differential operator of order k	7
E	fundamental solution for $\square$	7
P	Poisson operator	7
G	Green's operator	7
$B_{\bar{\partial}}$	boundary operator for second boundary condition	8
$\square^+$	$= B_{\bar{\partial}} P$	8

Symbol	Explanation	Chapter of first occurrence
$\square^-$	another operator similar to $\square^+$	8
K	parametrix for $\square_b$, $n > 1$	8
N_a	approximate Neumann operator	9
R	operator of remainder type	9
S_m	operator of Heisenberg-group type m	9
$\overline{K} = \overline{K}_u$	parametrix for $\square_b$ relative to $\overline{C}_b$, when $n=1$	10
Q_{E^-}	inverse of $\square^+$ away from characteristic variety	10
N	(exact) Neumann operator	11
$\mathcal{H}$	space of harmonic forms	11
B^p	Besov space	12
Λ_α	Lipschitz space	13
L^p_k	Sobolev space	14
Allowable vector fields		15
$U = \vartheta N(f)$	solution of $\bar\partial U = f$	16
$H(f)$	Henkin solution of $\bar\partial U = f$	17

References

0. Agmon, S., Douglis, A., and Nirenberg, L., "Estimates near the boundary for solutions of elliptic partial differential equations satisfying general boundary conditions, I and II," Comm. Pure and Applied Math. 12 (1959), 623-727; 17 (1964), 35-92.

1. Alt, W. "Hölderabshatzungen für Ableitungen von Lösungen der Gleichung $\bar{\partial}u = f$ bei Streng Pseudo-konvexen Rand," Manuscripta Math. 13 (1974), 381-414.

2. Boutet de Monvel, L. "Comportement d'un opérateur pseudo-differentiel sur une variété à bord I, II," J. Anal. Math. 17 (1966), 241-304.

3. __________, and Sjöstrand, J., "Sur la singularité des noyaux de Bergman et de Szegö," preprint.

4. Calderón, A.P., "Lebesgue spaces of differentiable functions and distributions," Proc. Symp. Pure Math., vol. 4, Amer. Math. Soc. (1961), 33-49.

5. __________, "Boundary value problems for elliptic equations," Joint Soviet-American Symposium on partial differential equations, Novosibirsk (1963), 303-304.

6. Erdélyi, A. etal., "Higher transcendental functions," vol. I, McGraw-Hill, New York, 1953.

7. Fefferman, C., "The Bergman kernel and biholomorphic mappings of pseudo-convex domains," Invent. Math. 26 (1974), 1-65.

8. Folland, G.B., and Kohn, J.J., "The Neumann problem for the Cauchy-Riemann complex," Ann. of Math. Studies #75, Princeton Univ. Press, Princeton, N.J., 1972.

9. Folland, G.B., and Stein, E.M., "Estimates for the $\bar{\partial}_b$ complex and analysis on the Heisenberg group," Comm. Pure and App. Math. 27 (1974), 429-522.

10. Grauert, H., and Lieb, I., "Das Ramirezsche Integral und die lösung der Gleichung $\bar{\partial}f = \alpha$ in Bereich der beschränkten Formen," Rice Univ. Studies 56, 2 (1970), 29-50.

11. Greiner, P.C., Kohn, J.J., and Stein, E.M., "Necessary and sufficient conditions for solvability of the Lewy equation," Proc. Nat. Acad. Sci. 72 (1975), 3287-3289.

12. Greiner, P.C., and Stein, E.M., "A parametrix for the $\bar{\partial}$-Neumann problem," Proceedings of the 1974 Montreal Conference "Rencontre sur l'analyse complexe à plusieurs variables et les systèmes indéterminés," Université de Montréal (1975), 49-63.

13. Henkin, G.M., "Integral representations of functions holomorphic in strictly pseudo-convex domains and applications to the $\bar{\partial}$ problem,' Mat. Sb. 82 (124) 1970, 300-308; Math. U.S.S.R. Sb. 11 (1970), 273-281.

14. ———, and Romanov, A.V., "Exact Hölder estimates of solutions of the $\bar{\partial}$ equation," Izv. Akad SSSR 35 (1971), 1171-1183; Math. U.S.S.R. Sb. 5 (1971), 1180-1192.

15. Henrich, C.J., "Characterization of the boundaries of complex manifolds I, The Lewy complex," (preprint).

16. Hirschman, I.I., "A convexity theorem for certain groups of transformations," J. d'Analyse Math. 2 (1953), 209-218.

17. Hörmander, L., "Linear partial differential operators," Springer, Berlin (1963).

18. ———, "Pseudo-differential operators and non-elliptic boundary problems," Ann. of Math. 83 (1966), 129-209.

19. ———, "Pseudo-differential operators and hypoelliptic equations," Proc. Symp. Pure Math. 10 (1967), 138-183.

20. Kerzman, N., "Hölder and L^p estimates for solutions of $\bar{\partial}u = f$ on strongly pseudo-convex domains," Comm. Pure and Appl. Math. 24 (1971), 301-379.

21. Kohn, J.J., "Harmonic integrals on strongly pseudo-convex manifolds I and II," Annals of Math. 78 (1963) 112-148; 79 (1964), 450-472.

22. Korányi, A., and Vági. S. "Singular integrals in homogeneous spaces and some problems in classical analysis," Ann. Scuola Norm. Sup. Pisa 25 (1971), 575-648.

23. Krantz, S.G., "Optimal Lipschitz and L^p estimates on strongly pseudo-convex domains," Princeton University Ph.D. Thesis 1974; "Optimal Lipschitz and L^p regularity for the equation $\bar{\partial}u = f$ on strongly pseudo-convex domains," Math. Ann. 219 (1976), 233-260.

24. ———, "Structure and interpolation theorems for certain Lipschitz spaces and estimates for the $\bar{\partial}$ equation," Duke Math. J. 43 (1976), 417-439.

25. Nikolskii, S.M., "Approximation of functions of several variables and imbedding theorems," Springer, Berlin, 1975.

26. Nirenberg, L., "Pseudo-differential operators," Proc. Sym. Pure Math. no. 16 (1970), 149-167.

27. Øverlid, N., "Pseudo-differential operators and the $\bar{\partial}$ equation," in Springer lecture notes, no 512 (1976), 185-192.

28. Phong, D.H., Ph.D. Thesis, submitted to Princeton University.

29. Rothschild, L.P., and Stein, E.M., "Hypoelliptic differential operators and nilpotent groups," Acta Mathematica (1976) (to appear).

30. Sato, M., Kawai, T., and Kashiwara, M., "Microfunctions and pseudo-differential equations," Lecture notes in Mathematics no. 287, Springer (1973), 265-529.

31. Seeley, R., "Singular integrals and boundary value problems," Amer. J. Math. 80 (1966), 781-809.

32. Siu, Y.T., "The $\bar{\partial}$ problem with uniform bounds on derivatives," Math. Ann. 207 (1974), 163-176.

33. Stein, E.M., "Singular integrals and differentiability properties of functions," Princeton, 1970.

34. ———, "Singular integrals and estimates for the Cauchy-Riemann equations," Bull. Amer. Math. Soc. 79 (1973), 440-445.

35. ———, "Singular integral operators and nilpotent groups," Proceedings of the C.I.M.E. Conference "Differential operators on manifolds," (1975), Edizioni Cremonese, 148-206.

36. Zygmund, A., "Trigonometric series," (2nd edition), Cambridge, 1959.

Library of Congress Cataloging in Publication Data

Greiner, Peter Charles, 1938-
Estimates for the $\bar{\partial}$-Neumann problem.

(Mathematical notes series ; 19)
Bibliography: p.
1. Neumann problem. I. Stein, Elias M., 1931- joint author. II. Title. III. Title: Delta bar-Neumann problem. IV. Series: Mathematical notes (Princeton, N.J.) ; 19.
QA374.G73 1977 515'.353 77-1687
ISBN 0-691-08013-5 pbk.